MICHAEL MORCOMBE'S

DISCOVERING BIRDS

BRISBANE & SURROUNDS

• Moreton Bay & Islands • Gold Coast & Hinterland
• The Scenic Rim • Sunshine Coast & Hinterland
• Fraser Island & Northern New South Wales

Steve Parish™
PUBLISHING

Contents

Introduction

South-eastern Queensland is one of Australia's richest areas for birdlife — not only in terms of diversity, but also for the spectacular beauty of many of the region's abundant species. The bird list for the Greater Brisbane area numbers just over 400 — representing about half of Australia's total bird species.

Torres Strait is narrow enough to allow many bird species to travel to or from Papua New Guinea. Birds whose home is essentially southern Australia also wander or migrate just far enough north to reach the Brisbane area, further adding to the local abundance. Other species are birds of the semiarid interior, which have increasingly moved to the south-east as land clearing has provided a new, greener refuge whenever the interior is exceptionally dry.

Migratory birds are another contribution to the bird population. Land birds, including many honeyeaters, monarchs, fantails, parrots and kingfishers, make annual migrations north in winter, returning south in summer. Most seem to pass through south-eastern Queensland, heading as far as north Queensland, Papua New Guinea and Indonesia.

Then there are the migratory shorebirds, or waders, which come to Australia in late spring from their breeding ground in the Northern Hemisphere and return north in autumn. The extensive mudflats and beaches of Moreton Bay's islands and channels form important feeding grounds for such visitors.

All this makes Brisbane a most fortunately placed city for bird life. However, the number of birds could not make use of the region were it not for Greater Brisbane's rich and varying habitats — rainforests, wet eucalypt forests, dry eucalypt forests and woodlands, heathlands, beaches, reefs and mangroves. These habitats, together with the important bird species that live there, are each represented in the chapters of this book.

There is also the urban habitat — artificial and comparatively new. Yet some species, being especially adaptable, not only tolerate such a habitat, but thrive alongside humans and urban progress. These birds make Brisbane's urban parks, gardens and backyards rich in bird species.

For those who are keen to know more about Brisbane's great variety of birds, the most rewarding way is to join one of the main bird organisations, or clubs. These cater for everyone from beginner to advanced birders. They offer excursions to many places where these birds can be best seen and can provide help with identification. Their websites and publications are immensely helpful. Contact information is provided at the back of this book.

Opposite: Familiar faces of Greater Brisbane — frolicking Galahs.
Above, top to bottom: Crimson Rosella; Eastern Yellow Robin.

Dry Eucalypt Forest Birds

Dry eucalypt forests and open woodlands cover a large part of the Greater Brisbane region, in both coastal and inland areas. These dry eucalypt forests are much more open than the region's wet eucalypt forests, but not as sparsely vegetated as those of drier country further inland than this Brisbane region.

Dry forests are widespread, usually occurring where soils are less fertile and rainfall lower. Eucalypts are dominant, but not as tall or close together as trees in wet eucalypt forests. The ground cover, consisting of shrubs or grass, is also relatively open. The forest canopy is indicative of the type of forest — as much as 70% cover in the heavy wet eucalypt forest, down to 10% for the more open woodlands. The transition from wet forest to dry forest to woodlands is continuous, so that the divide between dry forest and woodland can be vague.

In woodlands, the trees are even more widely spaced than in dry forests, although the transition from dry or open forest to woodland may be almost imperceptible. As with other forest types, "woodland" is defined in terms of canopy cover and tree proportions. The wider spacing of woodland trees gives a canopy cover of 10–50%. The trees have enough space and light to expand laterally and more of the lower limbs are retained than in forests. Within this broad description, dry eucalypt forests and woodlands vary greatly, ranging from grasslands with sparsely scattered trees, to forests of tall, closely spaced trees that approach wet forests in character.

Species such as the tallowwood and grey gum grow in more sheltered, moist locations. The spotted gum is common nearer the coast and on poorer soils. Where fires are frequent there is likely to be a ground cover of kangaroo grass, or perhaps blady grass. Elsewhere there may be a shrubby ground cover, with diverse flora of wattles, banksias, hakeas, grass trees and various smaller shrubs. In areas that are frequently waterlogged, especially over poor soils derived from sandstone, the forest understorey is made up of paperbarks, tea-trees, banksias and grass trees.

Many botanical names, applied at various times to distinctive types of vegetation, are used within the broad classifications of dry forest and woodlands. These may include some riverine forests, river red-gum and timbered country, paperbark woodland or swamp woodland. With all these variants, including those intermixed with areas of swamp and heath, the dry eucalypt forests and woodlands are, not surprisingly, home to a large number of bird species.

TOP BIRDWATCHING sites

DAISY HILL KOALA PARK

While the primary objective is to safeguard habitat for Brisbane's south-east Koala, it is of equal value for the protection of birdlife, woodlands, dry eucalypt forests and other habitats within the total 25,000 ha of this large conservation reserve.

GOOMBURRA STATE FOREST PARK

For Brisbane birdwatchers, this park, set in the western foothills of the Scenic Rim, provides an opportunity to see Eastern Rosellas, White-eared Honeyeaters and Musk Lorikeets. Another notable species is the Albert's Lyrebird.

GLASS HOUSE MOUNTAINS NATIONAL PARK

This series of eroded volcanic peaks is situated 75 km north of Brisbane off the Bruce Highway. Of special interest here are the high cliff ledges that provide nest sites for Peregrine Falcons.

BRISBANE FOREST PARK

An excellent woodland habitat within easy reach of the CBD. The entrance is at 60 Mt Nebo Road.

Above: Rainbow Bee-eater.

Rainbow Bee-eater *Merops ornatus*

name: Rainbow Bee-eater *Merops ornatus*
length: 23–27 cm
looks: body mostly vivid light green; undertail coverts pale blue; throat and eye stripe black; underwings orange
voice: sharp, cheery; in flight "pirr-pirr-pirr"; when perched, similar but softer
habitat: throughout Australia except Tasmania and most arid western sandy deserts
nesting: a tunnel drilled into flattish ground
abundance: common

Orange wings flash against the blue sky as Rainbow Bee-eaters twist and turn in pursuit of fast-flying insects. Their sharp shrill calls, sounded repeatedly, draw attention to their acrobatic flight and brilliant colours.

When Bee-eaters arrive from northern Australia, during October and November, they are noisy and active. They become much quieter as they disperse, each pair claiming a patch of open land, and settling in to drill their nesting tunnels.

These birds are most conspicuous in late summer, when they gather into flocks prior to departing for far northern Australia. At this time of year their numbers greatly increase as young birds, in plumage almost as spectacularly colourful as that of their parents, leave their underground nests. Then they join the colourful, raucous flocks that gather, before leaving for warmer northern regions where their flying insect prey will still be abundant even during the winter months.

With each pair laying up to seven eggs, the bee-eater population may be three or four times greater than it was on arrival a few months earlier — an increase that is spectacularly (and noisily) evident wherever the flocks are gathered. This may occur with birds flying low to the ground over open areas or elsewhere at heights where many birds can use the twigs of taller trees as perches between their fast bursts of flight.

Only the larger backyards of outer suburbs are likely to host resident bee-eaters. Here the soil needs to be soft enough to dig, but firm enough to retain the bee-eater's nest tunnel. Typically these birds prefer large mowed lawns and expanses free of shrubs, or an open paddock — spaces kept for a horse. These birds are wary of tall, dense grass or other vegetation that might conceal a cat, fox or other predator.

Unfortunately, there is not much else that can be done to attract bee-eaters as permanent residents. They are far more likely to be but passing visitors, negotiating routes from treetop to treetop, wheeling and swooping across the open spaces of suburbs and backyards.

Grey Butcherbird *Cracticus torquatus*

This species has long held the moniker butcherbird with good reason. Pound for pound they are among the deadliest predators of all bushland creatures and their grisly treatment of prey quickly moved European settlers to incorporate such behaviour into the species' common name. The name "butcherbird" also applies to three other species found in Australia.

Butcherbirds have small weak feet compared with the deadly talons of raptors. However, they do have a long, straight, finely hooked bill with which they are able to take quite large prey. Much of their hunting is directed at prey on the ground — the bird waits watchfully on a low limb, before executing a steep, fast, straight glide or dive to strike with the bill, or drops, landing and jabbing at prey on the ground.

Large insects and spiders form a major part of this bird's diet, but it also kills small reptiles, mice and small native mammals (in habitats where these still exist). Butcherbirds also hunt small birds such as finches (which feed on the ground), the eggs and nestlings of ground-nesting birds (such as pipits and songlarks), and probably almost any small bird of the shrubbery and foliage that can be attacked unawares.

The Grey Butcherbird can also dismember prey without using its feet. Instead, the body of its prey is often wedged into a narrow fork of a tree and held fast while the butcherbird rips it apart with its hooked bill. Food not immediately eaten may be left hanging in a tree fork.

The Grey Butcherbird's nest is a small bowl of sticks, lined with softer grasses, built into a tree fork at heights of 3–10 m above the ground. The female builds the nest and incubates the eggs, while the male vigorously defends the site. When the young have hatched, he helps with their feeding. Intruders on their domain are fiercely set upon by both adults.

The Grey Butcherbird is known for its rich and varied calls, including musical and mellow songs as well as harsher sounds, emphatic and far-reaching notes, piping sounds, and strident scoldings in defence of the nest.

name: Grey Butcherbird *Cracticus torquatus*
length: 26–30 cm
looks: rather like a small magpie; grey back, wings, tail; face is black with sharply down-pointed black cheek patches; rump white; tail tipped white; underbody white; wing stripe white
voice: varied; both harsh scoldings and mellow musical sounds; loud piping song in flight
habitat: diverse; rainforest margins through eucalypt forests and woodlands, paperbark swamps, drier inland scrubs
nesting: untidy-looking bowl of sticks, but with the nest cavity smoothly lined; placed in a tree fork 3–10 m above ground
abundance: common, sedentary or locally common

Dollarbird *Eurystomus orientalis*

name: Dollarbird *Eurystomus orientalis*
length: 25–29 cm
looks: usually noticed in its acrobatic aerial swooping and wheeling flight as it chases insects with white "dollar" wing patches on dark wings; when perched shows large brown head, pink bill, green wings and tail, tiny pink feet
voice: fast cackle, as an abrupt, grinding, harsh, buzzing, rattled "kzak-kzak-kzak"
habitat: woodlands, open forests, farmlands, edges of heavy forests, city suburbs
nesting: unlike the migrant shorebirds, the Dollarbird breeds in Australia from Oct–Feb; nest is high, but in a shallow hollow of a tree
abundance: summer migrant to Australia; common, more so in north than south-east

This bird is conspicuous and noisy early in its breeding season, performing diving and rolling display flights. It pursues insects in fast flight from high exposed perches propelled by deep, powerful wingbeats and brief glides.

Each outstretched wing displays a rounded, crescent-like, whitish patch across the base of the primary feathers, on both upper and lower surfaces. The rough likeness of this patch to a dollar coin gives the "Dollarbird" its name. On the very dark, almost black violet-blues and greens of the wings, these bluish-white areas are highly conspicuous.

When perched, the Dollarbird has a distinctive silhouette displaying a large head with little narrowing of the neck, and a body tapering down to a stumpy tail. It never comes to ground, even to walk or hop, but waits almost motionless between darting flights to take passing insects. All prey is taken in flight — the Dollarbird performing acrobatic aerial manoeuvres to follow and snap up larger flying insects such as cicadas, mantids and beetles in its broad bill. Most of this hunting is done very early in the morning and late into twilight.

A common sound from this bird is a rapid cackle, an abrupt, harsh, metallic buzzing "kzak-kzak-kzak" or a faster "kzak-kzak-kzak" or "kakakakakak". Other calls include a deeper growling "kzaal" and a raucous, but not grating, "kraak".

The Dollarbird uses the airspace over many habitats, mostly open country with scattered trees, including woodlands, the margins of forests, farmlands, and city suburbs. It is a summer migrant to Australia, and the only member of the tropical roller family to visit the country. Rollers have their greatest numbers in the African region, but are also found from India to Indonesia, deriving their name from their tumbling and rolling aerial acrobatics. All rollers are brightly coloured, especially noted are their bright blue wings.

The Dollarbird's nest is a high hollow of a tree. Its eggs are laid on the bare wood-dust bottom of the hollow and are probably incubated by both adults.

Double-barred Finch *Taeniopygia bichenovii*

This bird is sometimes described as the "Owl-faced Finch", which is indeed an appropriate description. The white face with narrow, encircling black band is quite like the face of a Barn Owl in miniature. It differs from the owl by its distinctive second black band across the breast, which gives this bird its name.

Double-barred Finches prefer grassy woodlands and open forests. Here they feed in flocks on the ground where drying grasses drop their seeds. The birds pull down overhanging grass stems to bring the seed heads within reach.

When disturbed, the flocks depart with undulating, rather bouncy flight, drawing attention with their brassy, drawn-out "tzeeeaat–tzeeeaat" contact calls, which are rather like those of the well-known Zebra Finch but higher, weaker and more plaintive.

Double-barred Finches prefer the better-watered grassy woodlands of northern and eastern Australia. The species seems to be expanding into coastal eastern Australia, benefiting from thinning and clearing of forests. Colonies of these finches use open grassy areas for feeding, dense low shrubbery for shelter, and permanent pools or streams where they need to drink daily. They are never far from water, visiting it frequently throughout the day.

These finches are highly social, often gathering into flocks of 40–50 birds. The flocks feed together and cram into roosting nests at night. While pairs stay together for much of the year, they remain part of a larger flock. Not until the spring–summer breeding season do flocks break into individual pairs for nesting. Their courtship displays are simple, without any elaborate dance or ritual. The male merely fluffs up his plumage, pivots from side to side on his perch, and sings a simple soft song.

The nest is built in a low shrub 1–4 m above ground. This is a rather rough sphere of grass with a side opening and short entry. It is warmly lined with fine grasses, feathers and plant down and will contain a clutch of four to seven white eggs.

name: Double-barred Finch *Taeniopygia bichenovii*
length: 10–11 cm
looks: owl-faced appearance in miniature; narrow encircling black band around face and across chin; second band across breast; underparts white; back pale brown; rump white on male, black on female; tail tip blackish
voice: similar to the well-known, widespread and common aviary bird, the Zebra Finch; brassy or nasal "tzeeeaat", each of series more extended
habitat: grassy-floored woodlands and open forests, grassy scrublands, farmlands; never far from water
nesting: nest is built in various places; in bush, palms, hollows, under eaves of buildings; nest is a rounded-domed construction with hooded entry
abundance: common, nomadic

Noisy Friarbird *Philemon corniculatus*

name: Noisy Friarbird *Philemon corniculatus*

length: 30–35 cm

looks: very large honeyeater; bare black head; high bump or knob on top of large downcurved bill; red eye; brown back; underparts pale buff; other similar species outside Greater Brisbane region

voice: noisy, especially around flowering trees; often in quarrelsome flocks; harasses and aggressively dominates other honeyeaters; call is a rather goose-like honking "owk-orrok, owk-orrok" and similar sounds

habitat: open forests and woodlands, heathlands, suburban gardens

nesting: builds a large open cup of bark strips and fine twigs; suspended in outer foliage, 2–5 m above ground

abundance: common

This very large, unmistakable honeyeater is attracted to nectar, insects and fruit, many of their kind gathering wherever there is an abundance of food to be exploited. Their rackety calls and noisy squabbling will continue as long as eucalypts or other flowering trees provide nectar, or there is fruit ripening.

Like other honeyeaters, Noisy Friarbirds feed in trees and shrubs, energetically hopping along the branches, jabbing into bark crevices, plucking ripe fruit and probing flowers for nectar. At times this species will tackle flying insects — the birds rising steeply above the foliage, then gliding back down into the trees.

The Noisy Friarbird's bare head and bill knob are the source of some conjecture. Possibly the bare head avoids the problem of plumes getting sticky with nectar. The knob could be used in courtship and the selection of mates, or its surface hollows could play a role in the storing and transfer of pollen.

Often many other species of honeyeater compete for the same food source, commonly Little Friarbirds, wattlebirds and small honeyeaters. In this situation the Noisy Friarbird will become particularly aggressive, issuing loud warning calls as well as harassing and intimidating other birds to defend its right to a particular feeding area.

The Noisy Friarbird's voice is a deep, rollicking, rather goose-like honking "owk-orrok, owk-orrok, arrowk-kok-korrowk", parts of which have been described as sounding like "tobacco", or "four o'clock".

The Noisy Friarbird is a conspicuous autumn migrant, with flocks of 30–50 birds travelling up the coast, passing through south-eastern Queensland and heading high and direct towards central and northern Queensland. From August to December they return to their south-eastern nesting territories, returning in small numbers, wandering through forests and woodlands.

Friarbirds use a variety of habitats including open forests, woodlands, mangroves and suburban gardens. Their nest territory may be located in any one of these habitats, with the nest built into the outer foliage of trees anywhere from 2–20 m above ground. The deep, cup-shaped nest is suspended from its rim, the walls woven from grasses and bark strips, lined with finer, softer materials.

Tawny Frogmouth *Podargus strigoides*

The Tawny Frogmouth is a nocturnal hunter. At dusk it stirs from daytime lethargy to find the small creatures which are its prey. Hunting is undertaken from a post or tree limb. Here the frogmouth waits patiently until some slight sound reveals a small creature on the ground. The bird then swoops silently upon its prey (generally a spider, mouse, gecko or moth), engulfing it within its massively wide bill. The feet of these birds are small, slender and weak (when compared to those of owls and the diurnal birds of prey), and play no part in the bird's kill.

Tawny Frogmouths are often seen perched on roadside fences. Probably the surface of the road, and other areas of bare ground, make the detection and capture of prey easier (when compared with ground that has grass, shrubbery, litter, or other concealing cover). On cool nights, dark-coloured sealed roads that have become hot during the day remain warm well into the night and attract reptiles and other creatures that need warmth to remain active.

By day this beautifully patterned bird is renowned for its extremely effective camouflage. Long before any intruder is near its roost or nest it has assumed its famed "broken branch" pose — body rigid, head thrust forward, its shape and colour presenting an amazing imitation of a broken branch.

The call of the Tawny Frogmouth is a resonant "oom-oom-oom", repeated many times. A common misconception is that frogmouths give a double-noted "mo-poke" call, but that call belongs to the Southern Boobook — a small, brownish owl that, like the Tawny Frogmouth, is found almost throughout Australia. Of course, there is little chance that either bird will be observed calling at night, so the legend persists, and Frogmouths are commonly known as "Mopokes".

The nest of the Tawny Frogmouth must surely be one of the worst examples of avian architecture. Flimsy, rough and untidy, it is but the most basic mesh of sticks balanced across the fork of a tree. At the first signs of any intruder, the Tawny Frogmouth will adopt its upright, "invisible" posture — effectively imitating a broken branch beside a loose collection of twigs that appears to have fallen and been caught in the nearby fork.

name: Tawny Frogmouth *Podargus strigoides*
length: 34–52 cm
looks: best-known of Australia's night birds; daytime "broken branch" posture unique and well known; very large, broad head and bill; eyes are closed narrow slits by day; at night they become wide open eyes; plumage grey, mottled, bark-like
voice: low, resonant "oom-oom" from males; slightly faster "aroom-aroom" from females
habitat: open forests, woodlands, farmlands, suburbs, less in heavy forests
nesting: very rough platform of sticks placed across a horizontal fork or shallow cavity of a tree
abundance: common, sedentary

Peregrine Falcon *Falco peregrinus*

name: Peregrine Falcon *Falco peregrinus*
length: 35–50 cm
looks: heavily-built falcon; black cap down over eyes to broad black point on lower cheek; yellow eye-ring; breast to undertail white to pale buff, with broken cross-barring from breast back; thick yellow talons; fast flight on backswept narrow-pointed wings
voice: calls mostly in breeding season; very noisy around nest; loud "kiek-kiek-kiek" and rapid, agitated "kek-kek-kek"
habitat: widespread; rainforest to arid scrublands, coastal heath to alpine; requires seclusion from excessive human disturbance where nesting and needs cliff or large tree hollows
nesting: uses a ledge on a cliff-face, or a high, wide, shallow hollow of a large tree; eggs are laid into a slight depression of rock surface or on bare wood of tree hollow floor
abundance: generally uncommon, scarce in some areas

The Peregrine Falcon is one of Australia's most spectacular raptors — a marvel of sheer power and speed in its aerial pursuit and attack of prey.

Typical of the falcons, it has backswept, narrow-pointed wings, and a smoothly contoured body. In flight this falcon is fast and agile, travelling swiftly with stiff, shallow wingbeats broken by brief glides. In pursuit of birds these wingbeats become deeper, powerful, lashing strokes. When soaring high, its profile changes; wings are held straight out, so that the trailing edge of each is almost one straight line, the tail fanned and rounded.

Most spectacular is its stoop — the bird plummeting downwards, bullet-shaped with wings closed to strike at tremendous speed. The momentum of its combined speed and weight, and the shock of the impact through heavy powerful talons, sends fleeing prey tumbling towards the ground. The falcon will follow and, if the victim is large, feed upon it on the ground. Smaller birds, such as parrots and pigeons are likely to be overtaken by the diving falcon and seized in flight.

The Peregrine Falcon mostly hunts birds around parrot size, but will also take far larger prey, including heron, ibis, black-cockatoos and ducks; at the other extreme, prey also includes pardalotes, honeyeaters and large insects.

The Peregrine Falcon is the only falcon of such size and appearance likely to be seen in south-eastern Queensland. The Australian Hobby is similar, but much smaller and lighter in build. If observed together, there are a number of differences between these species.

The Hobby's breast has dark brown vertical streaks, where the adult Peregrine Falcon has fine cross-barring. Both have a black crown, and a black cheek patch with a shape unique to each species. The black cheek of the Peregrine Falcon has, at its lowest part, a broad point of about a 90° angle, where the Hobby has a much finer, long, narrow point.

The Peregrine Falcon uses, where possible, a ledge of a cliff as a nest site, laying its eggs on the rock. If there are no cliffs, large hollows in tall trees are used and the eggs are laid on the wood of the tree hollow.

Australian Hobby *Falco longipennis*

This small, fast-flying falcon is darker but similar in size to the well-known Nankeen Kestrel. The obvious differences between the Kestrel and Hobby are in their flight and in the type of prey that is hunted by each species.

The Kestrel takes small prey from the ground (grasshoppers, crickets, small skinks, mice, small and nestling ground birds) — it only occasionally chases small birds in flight. When hunting is slow, it spends its time hovering in flight. By contrast the Hobby is an aggressive hunter, taking birds in swift aerial pursuit. These include small ground-feeding birds and birds of foliage and treetops, up to the size of doves and small parrots. Large insects, including dragonflies and grasshoppers, are also occasionally taken in flight.

This small falcon uses mostly open habitats, typically open woodlands, and stays in the vicinity of trees along watercourses of open plains, around lakes and rivers, and in urban areas. Its flight is usually rather leisurely, but when hunting the wingbeat is a rapid flickering, with flight very fast, often low, and marked by twisting, turning and abrupt changes of direction. The speed of the falcon and its aerial acrobatics allows it to check any desperate attempts by prey to reach a sheltering shrub.

The Hobby very rarely lands or takes prey on the ground — prey is almost always taken in the air and then to a perch or nest where it is plucked and eaten.

The Hobby's colour and the markings of its plumage might be confused with the Peregrine Falcon — a larger, more heavily built, and even more determined and aggressive hunter that takes birds up to the size of ducks, ibis and herons while it is in flight.

The Hobby's breeding season is always from spring to summer, with egg laying somewhere between July and November. The Australian Hobby does not build its own nest, but uses an old nest, usually of a crow — these typically being built in the highest part of the crown of a tall eucalypt. The female incubates the two to four eggs. The male hunts for both, as well as for the chicks until they are old enough to be left alone as both adults go off to hunt.

name: Australian Hobby *Falco longipennis*

length: 30–35 cm

looks: small falcon, similar size to Nankeen Kestrel but does not hover; upper surfaces grey; underparts pale to mid buff, streaked darker; face distinctive; black crown joining black face patch with narrow point aiming downwards

voice: sharp, metallic "kiek-kiek-kiek" accelerating to "ki-ki-ki-ki" when excited or agitated

habitat: open forests and woodlands, heathlands, farmlands, suburbs, especially around and over swamps, riverbank tree belts, watercourses

nesting: uses old nests of other species, usually crow or raven, or another bird of prey, typically among rather thin branches of highest part of a tall tree; nests are quite large bulky platforms of sticks

abundance: generally uncommon

Pale-headed Rosella *Platycercus adscitus*

name: Pale-headed Rosella *Platycercus adscitus*
length: 28–32 cm
looks: typical rosella, but head and neck is bright light yellow
voice: in flight, an abrupt "czik-czik"; from trees, soft high thin tremulous "feee-ee-feee-e-ee"
habitat: woodlands, open forests, farmlands, roadsides, parks and gardens; feeds on native and grain seeds and fruits
nesting: tree hollows, which may be quite small and high; at times low in a stump or hollow post
abundance: common, sedentary

The vivid plumage of the Pale-headed Rosella would seem to make this bird an easy target for predators. But observers of this species in its natural environment have commented that when feeding on sun-dappled ground — where there are dead leaves, pieces of sun-bleached dry wood, fresh yellow and green shoots of foliage and dark shadows — the camouflage of this bird is excellent. Perhaps these birds know this, for they seem to prefer ground feeding, keeping to the dappled shadows of foliage across the ground, rather than being exposed in the glare of full sunlight.

Similarly, when feeding or resting among the foliage of trees, the colours surrounding the bird match and intermingle with the similar bright whites, yellows and greens of the feathers, so that these colourful birds can, incredibly, blend effortlessly into their surroundings.

The Pale-headed Rosella's preferred habitats are grassy woodlands, farmlands with scattered trees and lines of trees along watercourses and roads. It also lives on dry, scrubby ridges but favours life in lowland areas rather than on the highest parts of ranges.

The Pale-headed Rosella is an abundant species and it is copious in certain areas. It seems that even in quite early times of European settlement, this parrot had a reputation as a prolific breeder. In 1909, collector H. G. Barnard marvelled at the fertility of this species (then known as the "Moreton Bay Rosella"). Barnard had taken clutches of eggs from the nests of Pale-headed Rosellas in the months of February, March, April, and from June to December. This species, he wrote, "breeds at almost any time of the year".

The records of early collectors provide details about this bird's nesting habits. Barnard's notes show he took eggs from a hole in a stump with an entry less than 2 m high. At the other extreme, he cut steps with an axe up the trunk of a dead tree to a nest entry some 22 m above ground in order to collect a clutch of eggs.

The near-continuous breeding cycle of this species and the typically large clutch of four to six eggs, helps to explain the Pale-headed Rosella's abundance.

Galah *Cacatua roseicapilla*

The Galah is so common in many parts of Australia that its beauty goes mostly unappreciated. If it was a rare species, the delicate soft pinks and greys of its plumage would always attract praise. The Galah's beautiful plumage, coupled with an entertaining personality, creates a most spectacular scene of colour and motion.

Often a flock that is feeding on the ground by the side of the road will explode into flight, turning the massed birds into a cloud of pink and grey. Wheeling overhead, they show the pale grey of their upper plumage against the blue of the sky. They reveal even more dramatically the bright pink of the underbody and wings when turning in unison. In sunlight the brightness of colour makes the sky seem darker, and the whole scene — the massed wheeling, turning and circling of pink and grey against deep blue — appears even more spectacular.

At other times Galahs put on a comic show, acting the clown with various silly antics. A favourite party trick seems to be hanging upside down from a branch by one foot, while flapping wings and screeching loudly, only to be ignored by the rest of the flock.

Galahs prefer an open environment — woodlands, farmlands and the outer suburbs where there are vacant lots, roadsides, and patches of open woodland. Their numbers have increased greatly since European settlement, with the clearing of heavier forests, and the provision of water in farm dams. There has been movement south and east into areas where the clearing of dense forests has created suitable open environments.

The Galah makes its nest in a large hollow of a tree, either in the trunk or in one of the major limbs. Typically, areas of bark have been removed on various parts of the tree, where the Galah's destructive energy and bill-sharpening have exposed the dead grey wood.

The clutch of two to five white eggs are laid on a lining of gum leaves at the bottom of the (typically very deep) hollow. Incubation takes 30 days, and the young are in the nest for about 42 days.

name: Galah *Cacatua roseicapilla*

length: 35– 38 cm

looks: well known pink-and-grey cockatoo; common cage bird; upperparts entirely soft, mid to pale greys; most of underbody and underwing coverts deep pink; crown of head has short pale crest

voice: rather harsh and metallic, yet not unpleasant "chirrink-chirrink, chirrink-chirrink"; loud harsh screeches when alarmed

habitat: diverse; mostly open country of dry forests, woodlands, farmlands, semiarid scrublands, outer suburbs

nesting: large, deep hollow of tree trunk or limb, lined with eucalypt leaves

abundance: very common in suitable habitat

Blue-faced Honeyeater *Entomyzon cyanotis*

name: Blue-faced Honeyeater *Entomyzon cyanotis*

length: 25–31 cm

looks: mid-sized honeyeater; unique large oval of bare skin around each bright yellow eye; surrounding plumage of head black; face, throat to central breast black; upper body and wings olive-green; underbody white

voice: noisy, with varied sounds, including a husky "kierrrk", repetitive contact call "whik-whik", and loud, clear, whistled "quorrieek"

habitat: open forests and woodlands of eucalypts or paperbarks, plantations, farmlands, mangrove edges, suburban gardens

nesting: sometimes breeds in small, loosely assembled colonies; nest is a rather bulky bowl constructed of bark strips in tree fork at heights of 3–20 m; may refurbish old babbler nest

abundance: common, locally nomadic

The unusual and conspicuous feature of this large honeyeater is the oval patch of bare skin around each eye, which is bright blue on adult birds, but intense greenish yellow on juvenile birds.

The vibrant colour of this patch is emphasised by the surrounding black plumage of the head, and the large black "teardrop" running from the bird's throat to its breast. The bright yellow eye, surrounded by blue, gives this honeyeater its remarkable appearance.

Gregarious Blue-faced Honeyeaters have communal nesting areas where, in a particularly favourable locality, up to ten or twelve birds will nest close together and forage in small working parties. They seek the nectar of flowering eucalypts, paperbarks, bottlebrushes or grevilleas, taking both wild and cultivated fruit, and hunting insects and other small prey living on tree foliage and bark. These birds are often seen climbing over branches, probing crevices, and often launch acrobatic aerial pursuits to capture large flying insects.

Blue-faced Honeyeaters are particularly innovative in their choice of nest site. Rather than build their own nests, this species will often take over and refurbish old nests of other birds. They appear to choose nests made by species that build large bulky structures of sticks. These nests are sturdier than those built of finer materials and are still useful long after their original owners have gone.

The nests of Grey-crowned Babblers, which are huge structures that remain intact many years after being abandoned, are most popular. The Blue-faced Honeyeater may also utilise Magpie-lark and Apostlebird old nests, both being large, deep, strong bowls of mud reinforced with grass. The Blue-faced Honeyeater can replace sufficient soft lining into the top of these nests to suit its own needs.

Faced with a lack of available squats, the Blue-faced Honeyeater is quite capable of building its own nest from scratch — a relatively large, strong bowl of sticks, bark and grass constructed in the fork of a tree. Their eggs are an attractive pinkish-buff, spotted and blotched with purplish-red and brown.

Forest Kingfisher *Todiramphus macleayii*

This kingfisher is most likely to be noticed in the breeding season, which is sometime during September and October around Brisbane. Each pair will select a nest site or return to their previous territory if the birds are returning migrants.

While some kingfishers may maintain residency, numbers of this species appear to increase in spring and summer, when migration southwards boosts the local population. This is the time that their presence on or near a backyard is most likely to be noticed. Males spend much of their time perched on exposed branches, calling almost continually.

The Forest Kingfisher's call is distinctive, quite unlike that of the other local "dry land" species, the Sacred Kingfisher. Its call has been described as high and rolling — a rapidly repeated series of high notes, which gradually slows in tempo. The female may respond with similar notes. There is also a "machine-gun" call, comprising short bursts of rapid rattling. An attack call is given when diving at a rival kingfisher or other intruder entering the Forest Kingfisher's nest territory.

During spring the presence of a pair of noisy Forest Kingfishers, in or near any Brisbane backyard, is hard to ignore. This is also the time to keep an eye open for their nest. Forest Kingfishers are most active and noisy when establishing their home territory. Their nest site soon becomes obvious as they work at excavating a tunnel and egg chamber. Later, with eggs or young in their nest, they become much more discreet.

These birds tend to return each year to the same site. Usually an arboreal termite nest is used, at a height of 5–12 m above ground; occasionally a natural hollow of a tree may be used. Most of the nest tunnel drilling seems to be done by the female. She belts her bill against the tree, gradually chipping away and raking out the hard clay-like material until the nest is complete with a wider nest chamber at the end of the tunnel.

Forest Kingfishers may visit any backyard on migration or as part of their wider hunting territory, but to have a backyard with an arboreal termite nest does seem to be the only way to tempt Forest Kingfishers into calling your backyard home.

name: Forest Kingfisher *Todiramphus macleayii*
length: 17–23 cm
looks: upperparts, including crown, royal blue; underparts clear white without buff tint; white collar broken at hind neck on female; in flight, conspicuous white wing patches
voice: sharp, rapid "ki-ki-ki-ki-ki"; voice faster than Sacred Kingfisher
habitat: open forests, woodlands, margins of swamps, farmlands; prey includes small lizards, frogs, insects
nesting: tunnel and chamber drilled into an arboreal termite nest
abundance: quite common

Masked Lapwing *Vanellus miles*

name: Masked Lapwing *Vanellus miles*
length: 35–39 cm
looks: large, tall lapwing; unique, extensive, fleshy yellow "wattles" on head and face; crown, back of neck to sides of breast, black; back and wings mid to light olive-brown; underbody white
voice: loud and sharply penetrating, especially when aggressive in defence of nest or young; as alarm or threat "karrak-karrak", rising in pitch and tempo with intruder's approach
habitat: various open, short-grassed sites, both natural and modified, often in vicinity swamp, lagoon or dam
nesting: slight depression in surface of ground; eggs well camouflaged and quite difficult to see
abundance: moderately common

Zealous defence of its nest has enabled the Masked Lapwing to maintain a presence even in the city suburbs, where this bird may be found in parks, on sports fields, vacant lots and anywhere that there is an open area of grass, or a grassy-floored woodland habitat.

Away from the confined urban environment, the Masked Lapwing is more likely to be found where there is wet grassland. Often this is the margin between a swamp, lake, river or dam, and its surrounding open woodland of sparsely scattered eucalypts, paperbarks or other vegetation. These birds, on the other hand, will also frequent open grassy areas of woodlands, pastured or ploughed paddocks with scattered trees, or stony plains where there is no water anywhere nearby.

Masked Lapwings have survived in both urban and natural habitats where many other species have gone. This is no doubt due to their overbearing protectiveness of nest and young. This protective characteristic is marked by steep, fast dives at intruders (however large or small), using sharp bony spikes or spurs that protrude midway along the leading edge of each wing, helping to enforce their wrath. Because of these spurs this species was known as the "Spur-winged Plover".

Most predators sensibly flee in great haste when faced by their loud "karrak-karrak" and sharper "karrik-karrik-kek-kek-kik-kikik"; the swishing rush of their dives; and their obvious intent to strike with wing spurs and defend their own young.

Masked Lapwings may be seen in pairs or small family parties in the breeding season; at other times they may be encountered in large flocks. These birds walk slowly, occasionally stabbing at small prey on the ground.

The nest is typically on open ground, well away from any large vegetation. A shallow hollow is scraped into the surface of bare ground, then sparsely lined with fine roots and dry grass, or left unlined. The three to four eggs are ochre-buff — heavily spotted and blotched with dark brown and chestnut.

The Masked Lapwing is found across eastern and northern Australia. Birds in the far north have larger yellow facial wattles but lack the black band down each side of the neck that is conspicuous on the east coast birds.

Mistletoebird *Dicaeum hirundinaceum*

The Mistletoebird would often go undetected but for its loud, high cheerful call, or the telltale flash of the male's iridescent blue-black and fiery scarlet plumage. Mostly this is a bird found in the higher foliage, where it is not easily spotted from some 10–20 m below. It is even less likely to be noticed where it feeds in clumps of mistletoe. Binoculars are required to appreciate the beauty of the male's plumage. Probably the best chance of a sighting is when the birds are returning frequently to the one spot and nest-building. This is usually quite high among foliage, but may occasionally be low down in a shrub.

The Mistletoebird darts from tree to tree, its flight fast and direct, issuing lively sharp calls and remaining hidden among the high foliage. The Mistletoebird voice is very loud, cheery and far-carrying. The call is a sharp "tieech, tieech, tee-wich, tee-wietch-tieewietch-teewieta". It also has a loud, sharp, rollicking "kinzee-kinzee, perweetc-perweeta-perweeta".

The Mistletoebird enjoys a very close, mutually dependent, relationship with the mistletoe plant. The parasitic plant grows from an extremely sticky seed, which, if it adheres to a host tree's live twig or branch (and there germinates), it is able to tap into the bark for sustenance and grow. The plant needs its seeds carried from tree to tree and placed on those branches where they might attach and grow — the Misletoebird performs this duty.

When perched, this bird has the habit of twisting from side to side and brushing against its perch, ensuring that some of the sticky seeds are wiped against it: there the seeds adhere and begin a new mistletoe plant. This species is likely to be seen in any suburb where there are trees with mistletoe growths. The Mistletoebird is nomadic, flying to localities where different species of mistletoe have ripe fruits.

The nest is one of the most beautifully crafted of all small nests. Tiny and pear-shaped, it hangs from its roof and has a side entrance. It is made of soft plant down bound with webs resulting in a soft felt-like material. The female builds the nest and incubates two or three eggs; the male contributes to the feeding of its young.

name: Mistletoebird *Dicaeum hirundinaceum*

length: 10–11 cm

looks: male upperparts, including head, wings and tail, steely blue-black; throat, breast and undertail coverts intense scarlet; belly white with vertical black streak; female is pale grey above; underparts creamy-white except for pink on undertail coverts

voice: far-carrying, strong, rather squeaky-sharp whistle "tiech-ti-witch-teewitch" and rollicking "kinzee-kinzee, perwita-perwita"

habitat: wherever mistletoe grows; dry and open forests, eucalypt woodlands, farmlands, suburban parks and gardens with trees

nesting: tiny, soft, domed nest, suspended by its roof from a twig of outer foliage; web and soft plant material gives a soft, felt-like feel

abundance: common, nomadic

Olive-backed Oriole *Oriolus sagittatus*

name: Olive-backed Oriole *Oriolus sagittatus*

length: 25–28 cm

looks: upperparts, head to rump, deep olive-grey; tail grey with white tips; underbody white, heavily streaked dark grey; bill salmon-pink; eye bright red

voice: rather musical, soft, clear, rollicking-bubbling call "orry-orry-oriole"

habitat: forests and woodlands, farmlands, occasionally rainforests

nesting: builds a deep, soft cup; larger version of typical suspended honeyeater nest; nest attached by rim between twigs of outer canopy foliage, 3–15 m above ground

abundance: more common in north; migrant to south-eastern Australia

The softly-coloured, grey-green streaked plumage allows this oriole to merge into treetop foliage, where it would often go unseen but for its bubbling, rather musical calls. While the male is distinctly greenish on its back, the female is pale olive-grey. Both are white beneath, with heavy blackish streaks. This, together with the adult's intense red eye, is a much better clue to their identity, rather than the nondescript greyish olive of the back for which they are named.

More than likely, it will be the calling of these birds that will bring them to attention, rather than any great flurry of activity. The call has a mellow, but crisp musical quality. With a rollicking "orry-orry-oriole", this bird clearly pronounces itself to be, indeed, an oriole. The song is a prolonged, rambling version of this call, often intermixed with snatches of mimicry of other bird calls.

This bird inhabits forests, woodlands and only occasionally, rainforests. In the southern parts of its range this oriole is migratory. Further north it is a resident throughout the year. It has been recorded breeding most months, but most frequently from September to January. The nest is a deep, rather untidy cup of bark strips, grass, wool, and leaves, bound with webs and suspended by its rim among leafy twigs on the outer canopy of a tree, at a height of 3–15 m.

It is common around Australia and is recorded in every month of the year, although there may be some north–south movement. Outside the breeding season these birds may be seen in small flocks.

Olive-backed Orioles are rather sedate foragers that work their way steadily through the foliage of the upper canopy of trees, seeking native fruits and berries. At times they are not adverse to including cultivated foods such as figs and mulberries as well. As this is a treetop foliage species, it will not matter what, if any, vegetation there is towards ground level — it does not descend. But if there is lawn, introduced or native shrubs, paddocks or sports grounds, and if there are ripe fruits or berries on offer, there could well be those pleasant "orry-orry-oriole" calls descending from among the treetop foliage.

Striated Pardalote *Pardalotus striatus*

Australia's four species of pardalote are among the smallest of the country's birds. Most widespread of these is the Striated Pardalote, which occurs almost throughout Australia. Not surprisingly, given the great distances and diversity of habitats of this bird, it has unique local variations of pattern and colour.

In the past, these variations or races were considered a distinct species with their own widely accepted common name. Those of the Greater Brisbane area and across far northern Australia, have a solid black crown and were known as the "Black-headed Pardalote". Striated Pardalotes of the eastern Queensland area have also lost their own common name, yet their plumage remains distinct when compared with the white-crowned Striated Pardalotes of inland and southern Australia.

All pardalotes forage in tree foliage, using their stubby little beaks to take small creatures, especially scale insects, from leaves. When close, one can hear the click of bills as they work over the leaves and flowers. Occasionally they can be seen with a mass of tiny pollen-covered insects crammed in their bills.

The Striated Pardalote is a common and well-known bird throughout most of the Greater Brisbane area. It occupies a great variety of habitats, from dense rainforests, tall wet eucalypt forests, to open forests and woodlands. In the Greater Brisbane area these tiny birds would be visitors to many, if not most gardens, but would no doubt be overlooked among leaves that are a similar size to the bird. Its loud "witta-witta" or "chip-chip" call, however, has become one of the city's more common background noises even if it is difficult to trace the tiny, leaf-sized makers of these calls.

The breeding season extends from June to December. The nest is a narrow, 3 cm–diameter tunnel dug into a vertical or steeply sloping earth bank; in other, distant parts of its range, a tree hollow is used. A globular bark nest with a side entrance is built in an enlarged chamber some 30–50 cm in from the entrance. Four white eggs are laid, and both sexes share the incubation of eggs and feeding of the young.

name: Striated Pardalote *Pardalotus striatus*
length: 9.5–11.5 cm
looks: tiny; short-billed; stumpy-tailed; stripes around folded wing; prominent brow line
voice: loud, repetitive and often repeated "witta-witta"
habitat: trees of rainforests, eucalypt forests, woodlands, arid scrubs; forages in foliage eating small creatures, especially lerps
nesting: domed bark and grass nest in small hollow of tree, in bank of earth, house roof space
abundance: common

Crested Pigeon *Ocyphaps lophotes*

name: Crested Pigeon *Ocyphaps lophotes*
length: 31–35cm
looks: small pigeon; tall upright crest rising from the crown; head delicate light blue-grey; upper body, wings and tail grey-brown; colourful iridescent speculum panel across the wing
voice: rather musical, abrupt "whoo"; starts softly then builds up; also a quick "whoop"
habitat: open woodlands, farmlands, roadside tree belts, farmlands; never very far from water
nesting: very frail, rough small platform of twigs; almost flat across the top
abundance: very widespread and common

This elegant pigeon first attracts attention by its distinctive flight — swift and direct, with brief bursts of whistling wingbeats alternating with silent, fast, level glides on wings held flat and stiffly straight. A final long glide and it usually lands conspicuously on an exposed branch, tipping forward for a few seconds with crest raised and tail uplifted, before standing tall and alert.

The Crested Pigeon is easily identified, even from a distance, by its high slender crest. Once assured there is no danger, it may then glide to the ground to feed. Birds of family parties or small flocks scurry about the bare and sparsely grassed ground picking up fallen seeds of grasses or crops.

At takeoff, it is the sound — the sudden loud clapping of its wings — that catches attention. Often takeoff is from the ground, where a typical small group of five or six birds may have been feeding. When startled they will all run together, then take off with an audible clatter of wings.

The wing-clap sounds are also a feature of courtship, for example, during display flight the male pigeon rises very steeply from his perch to a height of 20–30 m with powerful noisy wing-flapping, then glides steeply down. On the perch, where he can be easily seen, the male's display includes bowing with his tail held vertically and widely fanned.

Although the pigeon's plumage is of softly muted blue-grey and grey-brown, the iridescent colours of the speculum, across the centre of the folded wing, vary with the direction of sunlight, changing through green, violet and gold with each slight movement of the bird.

The call of the Crested Pigeon is a musical "whoo". A series of these sounds starts softly, then become a louder, "whoo-whoo-whoo", and include a quick "whoop".

Crested Pigeons build a flimsy small platform of sticks among foliage at heights usually 3–6 m above the ground. Two white eggs appear, making up the usual clutch. Both sexes share incubation of the eggs and the feeding of the nestlings.

Grey Shrike-thrush *Colluricincla harmonica*

It could be said that the most accomplished songsters are those birds which have the plainest plumage. With this in mind, the Grey Shrike-thrush would be one of the best examples. As its name indicates, grey dominates in various tints from near-white through pale and mid-grey to dark grey on the wings and tail. The only noticeable exception is the dull brown colouring on its back.

In striking contrast to its dull plumage is the unique "colour" of the voice for which this species is renowned. The rich and varied repertoire of calls and songs includes many sequences of notes that are regularly repeated and easily recognised as belonging to this bird. Often the Grey Shrike-thrush will improvise, adding a few notes and embellishments to its usual arrangements. Its singing is performed with such perfection, that the result seems more than sufficient to compensate for its dull plumage.

There are high, clear and loudly ringing whistles intermingled with deeper mellow musical notes, and rich bubbly sounds. A common sequence is "quorra-quorra, WHIET–chiew", beginning with mellow throaty sounds and finishing with a clear high ringing whistle. This song is repeated many times and extended, modified and varied. The Rufous Whistler probably has a similar quality to its voice, but while it is a superb songster, it does not have the depth and variety of the shrike-thrush song. Its calls include a high whistled "whee-it" and "whiet-wheit-wheeeeit".

The Grey Shrike-thrush forages on tree trunks, limbs and through the foliage, finding insects and other small prey that occasionally includes the nestlings of other birds.

It typically uses an enclosed site for its nest, but not a deep hollow. Adequately protected sites may be provided by a deep fork between large limbs, the centre of a very dense clump of vegetation or, occasionally, a hole in a cliff-face. Here it builds an open cup-shaped nest of strips of bark, lined with fine grasses. The usual clutch is of two to four eggs, which the male helps to incubate.

name: Grey Shrike-thrush *Colluricincla harmonica*
length: 22–25 cm
looks: plain grey except olive-brown tint on back; rufous undertail coverts; whitish patch between bill and eye
voice: rich and varied; clear and ringing notes intermingled with deep mellow sound; one of Australia's best songsters
habitat: diverse; forests, woodlands, arid scrublands, gardens
nesting: thick-walled bowl of twigs and bark, lined with grass; nest in shallow hollow of tree, cliff, or in dense tangle of foliage
abundance: common

Bush Stone-curlew *Burhinus grallarius*

name: Bush Stone-curlew *Burhinus grallarius*

length: 55–60 cm

looks: tall, stately bird; grey, black and white patterned plumage, similar to patterns of dry woodland leaf litter; moves almost in slow motion; if disturbed it crouches low, or slinks away slowly and unnoticed

voice: far-carrying, wailing calls at night; eerie, drawn-out, mournful sound begins softly then rises and strengthens "wee-ier, wee-ier, whiee-eeir"

habitat: open woodlands, dry forests, where quite open at ground level, with only sparse shrubs, ground cover of low sparse grass, or ground litter of dead twigs and leaves between sparse low green plants

nesting: may be a shallow depression scraped into ground surface, usually among or beside a few rocks, logs; eggs are cinnamon with heavy spotting of brown and blend well into ground colours and textures

abundance: moderately common

The wailing calls of the Bush Stone-curlew are one of the characteristic sounds of the Australian bush at night, where these birds still survive. Across much of southern Australia these birds are now only to be heard in places far removed from the cities. Brisbane, however, is fortunate enough to have this bird not only in its outer surrounds, but in residential suburbs.

The call, however, carries far, particularly on a still night, and this species is so much more likely to be heard than seen. Each call begins quite low and mournful, becoming louder, higher in pitch, then fades away — an eerie, spine-tingling wailing "wee-ier, wee-ier, wee-ieer, whee-ieer-loo". Often the call is started by one bird, but others will join in so that the volume of sound grows until it sounds like this ghostly chorus is bearing down on anyone within earshot.

Being a master of camouflage, this unusual species easily avoids detection. By day, the Bush Stone-curlew's camouflage makes its difficult to find, despite being quite a large bird. Not only do the colours and markings of the plumage cause this species to merge with background detail, its movements and postures (which include freezing and sly skulking) contribute to its ability to avoid recognition.

Well before being sighted, the Bush Stone-curlew will have adopted a pose that will make it hard to recognise. It might move away, with stealthy movements as if in slow motion, probably holding its head low, or it may lie flat on the ground with its neck extended. To fly would be a last resort; only if its cover is completely blown and it is actually chased, will this bird take to the air. When flying, the Bush Stone-curlew uses quick beats of long stiffly held, downcurved wings. If the upper surface is visible, white patches towards the outer part of the black flight feathers can be seen.

Bush Stone-curlews nest on the ground, scraping out a very small shallow hollow in the earth, which may be beside a low bush, log or rock, or out on an open plain or field. The two eggs are pale brown, spotted and blotched with dark brown and black. They look similar to the surrounding ground and are very difficult to detect.

White-winged Triller *Lalage tricolor*

The White-winged Triller is a conspicuous, well-known migrant to southern Australia. It spends the autumn and winter months in the far north and returns between mid-spring to early summer. The trillers may not return to the same sites each year, but rather to localities where seasonal conditions are best.

From his perch, the male White-winged Triller dominates the woodlands with an almost ceaseless loud trilling during mid-spring and into early summer. In display flights too, where he flies high above his nest territory, wings beating rapidly, the male will proudly issue a vigorous rollicking trill.

The chattering trilling of the male White-winged Triller carries far through woodland where midday heat has slowed and silenced many other birds — but not the display flights and calling of this species. The overall effect is cheerful, the tempo fast. The call has an undulating pitch and intensity of notes "cwipa-wipa-wipa-wipa-wipa", "chiffa-chiffa-chiffa". He often switches to a slightly softer, but still rapid "chif-chif-chif-chif".

Trillers take a wide variety of food, including insects, spiders and some fruits and seeds. They forage over tree trunks, branches and foliage, as well as taking some prey in flight. Much hunting is done on the ground, fossicking on and about logs, rocks and in grass.

These birds often breed in a semi-colonial manner with several pairs nesting quite close to each other, but each pair has a large territory for foraging use. They are recorded breeding most months of the year, but most usually from September to December.

The nest, jointly built by the male and female, is a neat, shallow bowl, which is very small for the size of the bird. It is made of fine rootlets and grasses thickly bound and covered with webs. The nest is neatly blended with webs onto the thick fork in which it is typically placed, so that from below it appears but a thicker part of the branch. Although the usual site is in the open, away from screening foliage, the nest can be very hard to detect by anyone other than its makers.

name: White-winged Triller *Lalage tricolor*

length: 16–19 cm

looks: small neatly plumed black and white bird, rather like a sleek, slender Willie Wagtail, but without that bird's spreading or wagging tail; male has black upperparts; white patterning on wings; underparts all white; female brownish above, merging to buff beneath, but with some stronger patterning on wings

voice: vigorous, sustained, trilling chatter; far-carrying, intermingled sharp and mellow sounds "chiffa-chiffa-chiffa, chwipa-chwipa"

habitat: open forests, woodlands, suburbs and gardens

nesting: small, neat cup blended onto thick tree fork; from side nest resembles thickened bump on wood; nest made of grass and rootlets, thickly bound with webs; rather like Willie Wagtail nest, but much lower, flatter, and harder to see

abundance: moderately common

Wetland Birds

Freshwater wetlands encompass a range of habitats, from shallows of a few centimetres to storage dams of great depth. Each kind of water body has its own kind of vegetation, forming shelter and feeding territory for different birds.

Brisbane's wetlands can be grouped into perhaps four types — the clear streams tumbling over falls and into pools from high rainforests; the shallow reed beds and muddy swamps of floodplains (and within these the many farm and "backyard" dams); the few large, deep lakes and major reservoirs; the swamps of the low-lying wallum heathland plains, and the lakes of the coastal dunes of the mainland and islands. Some of the best examples of sand-lake wetlands are within Cooloola, southern part of the Sandy National Park, where lakes, sand dunes and brackish water attract many birds including pelicans and sea-eagles.

Important among Brisbane's wetlands are the many creeks that flow down from the encircling ranges. Mt Barney National Park's ranges may be only narrow, but with their encircling dense vegetation, they are home to many birds including Scarlet Honeyeaters and Azure Kingfishers.

The Brisbane River forms a valuable wetlands resource. It meanders across the region's flat plains, most of its length lined with trees that overhang the water, where Darter, various cormorants, White-necked Heron, Rufous Night Heron and other birds can find a perch. Other more secretive birds skulk along banks under overhanging vegetation, seeking various aquatic prey. Out in full sunlight, Darters and Great Cormorants hold waterlogged wings out to dry.

Meandering westward, inland from ocean, port and city, the Brisbane River receives tributaries. Some are large, but many are very small creeks. Where there are deeply sheltered channels beneath overhanging trees, dense thickets and overgrowing grass or reed, habitat is provided for another spectrum of birds — those preferring shadow places to hide in, or sneak after their prey.

The river floodplain would long ago have carried lowland rainforest vegetation, which now exists only as small remnants. Ideal for the Azure Kingfisher, there it hunts from a perch in the shadows near creeks. Floodplain areas and low catchments with reed beds, tall grass or dense lantana thickets, provide home for cisticolas, Australian Brush-turkeys and Pheasant Coucals.

As the main channel of the Brisbane River turns north, it is blocked by a dam that creates Lake Wivenhoe. Its storage waters provide a type of wetland habitat suitable for another group of birds, which prefer the open-surfaced waters of the dam.

TOP BIRDWATCHING sites

BOONDALL WETLANDS

A large wetland area on the mainland side of Moreton Bay, traversed by the Gateway Motorway, and occupying most of the low-lying wetland between Nudgee Beach and Shorncliffe. The wetland area is traversed by Nundah Creek, which meanders through from the south-west, and Cabbage Tree Creek, which forms the northern boundary. A circular walking track leads through the wetlands.

ENOGGERA RESERVOIR, BRISBANE FOREST PARK

The Araucaria Walk follows the shoreline of the Enoggera Dam for a distance of 5 km, and which can be most rewarding for bird sightings over the dam waters and in the shoreline forest.

NOOSA BIRD TRAIL

This trail features 32 sites situated in varying habitats throughout Noosa Shire. One of these is the Fearnley Bird Hide at Lake Macdonald. This lake and the nearby water treatment plant are excellent birdwatching sites and support species such as the Great Crested Grebe and Green Pygmy-Goose.

Opposite: Black Swans.
Above: Eurasian Coots.

Darter *Anhinga melanogaster*

name: Darter *Anhinga melanogaster*
length: 85–90 cm
looks: known as "Snake-bird" for its long sinuous neck and small head, especially snake-like when swimming body submerged and head only showing; plumage mostly black, with fine white streaking over much of plumage; adult male has chestnut on fore-neck; yellow bill and face; bold white streak behind eye; female similar; lacks chestnut on fore-neck; white underbody, greyer upperbody; in flight, male all black; female white beneath; flies with neck outstretched
voice: loud, harsh, penetrating, far-carrying "kar-kar-ka-k-ka-kakaka" and faster
habitat: wetlands both fresh and brackish, with logs in water, tree-lined banks; also in sheltered coastal estuaries, lagoons
nesting: breeds in colonies or as solitary pairs; builds large, rough platform of sticks, lined with leaves on limb overhanging water
abundance: common in suitable wetlands habitats

The Darter got its alternative name of "Snake-bird" for good reason. This likeness is seen as most appropriate when the bird is swimming, its body submerged and only its head and neck above water. The Darter's long sinuous neck curves and kinks as it moves its head up, forward and down (as it does when disturbed), showing a small slender head that merges, snakelike, where it joins the neck.

From a distance the Darter's bill (being straight and neatly in line with head and neck) may not be recognised as easily as it would when the bird is closer or viewed through binoculars. The swimming snake likeness may be quite convincing for those who are not familiar with the Darter's habit of swimming with its body fully submerged, when fishing or alarmed.

The best opportunity to study the Darter is when it is perched with wings spread out to dry. In this situation, the bird is reluctant to fly or to re-enter the water and may allow a slightly closer approach. Generally, however, these are wary birds, and it is best to study details from a distance with binoculars.

Like the cormorants, these birds do not have waterproof plumage. This enables them to dive deeply with less effort as water displaces any air that would otherwise remain trapped beneath the surface of feathers. There is, however, always the need to dry the wings before flying. This is done on a bare tree limb, usually out over the water, and is quite conspicuous.

Darters are superbly adapted to fishing. They prefer smooth-surfaced lakes, swamps or very calm sheltered coastal waters. When diving, they hunt quite deep and impale their fish, yabbies or small tortoises with a spearing jab of the straight bill.

The Darter is distinctive in flight, with its neck held out straight and long tail trailing back with wings held stiffly out from the body. Often gliding in a "flying cross" shape, Darters will soar high in this flying position.

The Darter's nest is a rough platform of sticks positioned on the fork of a limb reaching out over water on a quiet backwater of a river or lake.

White-necked Heron *Ardea pacifica*

The White-necked Heron is unique in the Brisbane region for its two-toned plumage; all other local herons and egrets are either entirely white, or entirely grey. Only one other Australian heron has a similar two-toned plumage — the far smaller Pied Heron of tropical northern Australia, which is rarely, if ever, seen this far south.

In flight this heron holds its neck folded back, like all other herons and egrets, but is distinguished by the bright white spots about midway along the leading edge of each wing.

The White-necked Heron feeds mostly on small creatures in shallow waters and pastures. It hunts by walking very slowly forward, watchfully. Occasionally it will resort to stirring muddy or weedy water with one foot, then it will stand motionless awaiting any movement from the small creatures it may have disturbed.

White-necked Herons usually live in isolated pairs in the southern parts of their range, more often in colonies in the north. The nest is a rough platform of sticks set in a tall tree.

name: White-necked Heron *Ardea pacifica*
length: 16 to 75 cm
looks: large heron, back and wings grey, neck and head white, usually has several rows of dark spots down front of neck, legs grey-brown. In flight shows unique conspicuous white spots midway along the leading edge of the wing.
voice: in alarm it gives a croak or two when taking flight; other calls at nest
habitat: mainly freshwater shallow wetlands, swamps, lake margins, shallows of rivers, wet grasslands. Occasionally coastal mudflats
nesting: a large rough platform of sticks in tree branch hanging out over water. May nest as solitary pairs, or in colonies with various other waterbirds
abundance: common, disperses to flooded inland areas, later returns with young, increasing local population

White-faced Heron *Ardea novaehollandiae*

This species is the most common of all herons — one or two can be seen on almost every small pool, swamp, dam, or flooded pasture. Small backyard goldfish and lily ponds are also likely to have an occasional visit from a White-faced Heron. These birds are not only seen on freshwater wetlands, but hunt actively on coastal estuary, mudflat, reef and mangrove habitats as well.

The White-faced Heron is named for its most distinguishing plumage feature. Its face, from forehead to throat is white, while the rest of the plumage is grey. During the breeding season there are some brownish-grey plumes running down the lower neck onto upper breast.

The only sounds from this bird are various grunts — upon landing, in alarm, at the nest and in courtship displays. Unlike many other waterbirds, this species remains in isolated pairs throughout their breeding season.

Nests are rough stick platforms, at times far from water. The three to five eggs of the usual clutch are pale blue.

name: White-faced Heron *Ardea novaehollandiae*
length: 66–69 cm
looks: small slender heron; pale to mid-grey, except forehead-face-chin white; rufous-grey plumes lower neck to breast in breeding season; legs yellow
voice: usually silent except in alarm and at nest, gives grunting and croaking sounds
habitat: diverse wetlands, including swamplands, wet pastures, margins of lakes, farms, mangroves; feeds on small prey, frogs, mice, large insects, fish
nesting: as solitary pairs, unusually in small colonies; builds a rough stick nest usually quite high in a tree, often well away from water
abundance: very common

Intermediate Egret *Ardea intermedia*

name: Intermediate Egret *Ardea intermedia* length: 55–70 cm
looks: smaller and lighter build than Great Egret, but more elegant; neck shorter; smoother bends without such pronounced and obvious kinks; corner of gape (mouth) not extending to rear of centre of eye; rather flat crown of head; bill yellow; legs blackish or part yellow
voice: silent away from breeding colony; deep croaked alarm call if disturbed at nest
habitat: freshwater wetlands, lakeside shallows, swamps; occasionally marine habitats including mudflats and mangroves
nesting: in colonies with other egrets, spoonbills, cormorants; nest is a rough, shallow platform of sticks, lined with a few leaves, built into trees over water
abundance: common, especially in north

The Intermediate Egret is most attractive in the breeding season when carrying long lacy plumes on its back and breast. The long fine filaments cascade out over the folded wings and tail. Enhancing this plumage, the normally yellow bill and yellow parts of the legs flush deep red, while the yellow facial skin changes to green.

The flight is similar to the Great Egret's, but the wings beat faster, while the shorter legs do not trail as far behind the tail-tip. It often travels in flocks, but in random groups rather than in formal straight or "V" lines.

This bird uses a variety of wetland habitats, including the shallows of lakes, swamps and billabongs. At times it can be found among mangroves and on tidal mudflats.

With a breeding season from October to March, it nests in trees over water in colonies, together with other species of waterbird. Nests are shallow bowls of sticks, lined with a few leaves. The usual clutch is four to five eggs and the incubation is shared by both parents.

Great Egret *Ardea alba*

name: Great Egret *Ardea alba* length: 85–105 cm
looks: largest by far of all-white egrets; outstretched neck longer than rest of body and tail; neck has obvious "kinks"; gape extends to rear of eye; bill and legs all or partly yellow
voice: usually silent, but gives deep croak when alarmed, a loud "argh-aargh"; low rasping croaks at nest
habitat: fresh and saline wetlands, including swamps, dams, riverbanks, temporary floodwaters, estuaries, mangroves, mudflats
nesting: in colonies with other water birds; nest large, rough, shallow platform of sticks, built among upper branches of trees standing in water
abundance: common and widespread

Compared to the Intermediate and Little Egret, the Great Egret has a much longer neck with much more obvious "kinks" along it. One of Australia's most stately birds, in flight it has slow deliberate wingbeats — its head pulled back to the shoulders, neck folded under and long legs trailing well beyond the tail-tip. Hunts with slow, stealthy movements, often pausing to stand motionless, staring intently into the shallows, neck extended, or folded back, and ready to jab or spear.

Long plumes appear on the lower back of adults in the breeding season (November to February) when the bill becomes entirely black. Later the plumes are lost and the bill returns to deep yellow except for a blackish tip.

The voice is a variety of croakings. If disturbed at the nest it takes flight with a guttural "grok grok".

The Great Egret nests in colonies, which can comprise of other egrets, herons, ibis, spoonbills, darters or other species depending on the location.

Little Egret *Ardea (Egretta) garzetta*

Compared with the larger egrets, this species has a lighter, more delicate build. Its flight is lighter, more buoyant, and not interrupted by gliding. Its behaviour is also useful in its identification.

Egrets tend to stand still waiting for some item of prey to move, but the Little Egret will also chase small fish in the shallows, and has been reported hovering over fish and plunging down to take them. Other hunting tactics include flicking out its wings (to startle prey in clear water) or standing on one foot and putting the other forward to stir the water ahead (to flush out prey from reeds). This egret has black legs, but the soles of its feet are yellow, so there is speculation that this brighter colour may either attract or frighten the fish which the egret wants to see.

The Little Egrets nest in colonies, along with other egrets, herons, spoonbills and cormorants, building rough, shallow platforms of sticks in clumps of trees or shrubs standing in the water.

name: Little Egret *Ardea (Egretta) garzetta*
length: 55–65 cm
looks: compared with other egrets, delicate, slender, sleek build; quicker, lighter movements; dashes about chasing small water creatures; bill and legs black
voice: usually silent; croak if suddenly alarmed, quite noisy in colonies with croaking and gurgling noises
habitat: freshwater and marine wetlands; forages around shallows of swamps, mangroves, mudflats
nesting: in colonies, which may be in clumps of trees in freshwater or saltwater wetlands, including mangroves; nest is a rough platform of sticks, with shallow depression on top for eggs
abundance: uncommon

Cattle Egret *Ardea ibis*

This small, squat egret colonised northern Australia in the 1940s and is closely associated with cattle through Africa and South-East Asia.

The species successfully forages in close company with grazing cattle, which flush out insects and other small prey. These egrets have profited from improved grazing lands and irrigation methods that support not only the cattle, but also prey for the Cattle Egrets. Behind fences and among valuable large grazing animals, they are not likely to be in danger, and are probably regarded by property owners as a useful natural tool for reducing the numbers of pasture-damaging grasshoppers and the like.

Cattle Egrets are highly sociable, living in small groups to very large flocks. This continues during breeding, which occurs in swampland colonies in company with various other waterbirds. Nests are clustered over clumps of trees in swamps. Each nest is a rough platform of sticks.

name: Cattle Egret *Ardea ibis*
length: 48–53 cm
looks: small stocky egret, almost always seen in flocks in paddocks with cattle; hunched posture, neck folded until spearing prey; in breeding plumage has orange-buff plumes from head down back and breast
voice: usually silent; deep "krok-krok-krok" in breeding colonies
habitat: pastures, croplands, tidal mudflats
nesting: in colonies, stick nests in trees in swamplands
abundance: common in northern Australia, expanding southwards

Little Grassbird *Megalurus gramineus*

name: Little Grassbird Megalurus gramineus
length: 13–15 cm
looks: small bird of dense reed beds and similar habitats; upperparts of plumage grey-brown, heavily streaked with dark brown, black and white; underparts off-white, with fine short dark streaks; rump dull rufous; forehead rufous; tail dark, quite long, tapers to a point
voice: mournful whistle of three notes; first low and soft, others higher, longer
habitat: wetlands with dense vegetation of reeds, cane grass and mangroves
nesting: small, deep cup placed in very dense clump of vegetation, usually reeds, where densely packed nature of vegetation supports nest and prevents it sliding down; inserted into the rim are a few wide, curved feathers which arch over the open top of nest cup, screening it from above
abundance: locally common

A rather mournful soft call, of three notes, is often the first indication of the presence of the Little Grassbird. Once the call is known, the bird may be detected in many swampy or marshy sites, and with some perseverance it will probably be seen climbing the stems of reeds or fluttering low through the open spaces between clumps of reeds.

The grassbird is most likely to be seen alone, occasionally as a pair and rarely a small family group. The dense swampy vegetation of its wetland habitat, combined with the bird's camouflage plumage, makes any sighting quite difficult, and lengthy observations of its habits are rare. It usually creeps, mouse-like, low through the dense vegetation. However, it seems to be curious, and may climb up higher to inspect any unusual presence.

The Little Grassbird may also be glimpsed flying in or out of a clump of vegetation, but most flights are short, between clumps of reeds or other dense swampland vegetation, and usually only after being flushed out by the observer or some other creature. The flight looks weak and fluttering, with the tail trailing low.

The first note of the call is brief, soft and low; the second and third notes last much longer, both drawn-out and plaintive with a higher unwavering pitch — "weep-whieee-whieeee". This call is repeated constantly for long periods, advertising the bird's presence and status as a territory holder. The alarm call, is a harsh chatter issued when the bird is surprised or when its territory is invaded.

The Little Grassbird is a secretive, skulking inhabitant of dense wet vegetation. It may show itself briefly when first alerted, but usually quickly flutters away, tail trailing and often partly spread. A brief glimpse may show it to be a rusty brown above with the pattern of some feathers accentuated by white margins. The crown is more brightly rufous, with black streaks, while the underparts are creamy white with dark streaks.

The usual nesting season of the Little Grassbird is August to January, but some may begin breeding as early as July. The nest is a very deep cosy cup of loosely interwoven grasses and leaves, and is well lined with feathers. These include feathers of larger swamp birds — some inserted vertically around the rim with their natural curved shape forming a domed cover over the nest.

Swamp Harrier *Circus approximans*

Usually the dark shape of this raptor will be seen flying low over swamp reed beds or around the shoreline of lakes. Occasionally it rises much higher in flight, then glides down to fly just below the full height of the reeds, showing only momentary glimpses of its wingtips. The harrier's distinctive hunting technique is to float slowly on long broad wings, just brushing the reed tops almost at water level between clumps of reeds. Coming suddenly close over any potential prey that might be in an open water channel or in the mud between the safe havens of dense reed beds, the harrier has only to drop a metre or so for its victim to be within reach of grasping talons at the end of long dangling legs.

Often the sudden appearance of the dark shadow so close overhead will cause small creatures to panic and run or fly across open water or ground, when it would have been better to freeze and stay put. Prey are taken by surprise by the bird's low beating over the top of any vegetation. The harrier hunts this way for small ducklings, frogs, reptiles, fish or small mammals such as rats.

At other times a Swamp Harrier will fly low over large flocks of waterfowl waders that gather on open water, or over waders that feed across open mudflats putting the flocks to flight, but seldom showing enough speed or agility to capture those swift and evasive birds.

The call of the Swamp Harrier is usually just a threatening chatter in quarrelsome competition for food. In the breeding season, while soaring high over the nest territory, a brief whistled "ki-ooo" is sometimes made.

The Swamp Harrier's nest is a wide platform of trampled-down reeds, surrounded by water up to perhaps a metre deep. The mass of vegetation touches the mud bottom or semi-floats with some support from reeds beneath. At other times the Swamp Harrier may nest on dry land in crops, heath or tall grass. Of the two to five eggs, usually only one or two of the chicks survive.

name: Swamp Harrier *Circus approximans*

length: 50–60 cm

looks: low-flying gliding over lakes and swamps with wingtips upswept; mid brown on back; whole underside pale streaked and barred brown; rump white; juveniles are almost overall dark brown with a pale rufous rump

voice: only rarely a chattering "kik-kik-kik", perhaps threatening only when competing for food; some other calls near nest

habitat: swamps, lakes, mangroves, some marine waters, heathlands, temporary floodwaters, occasionally dry land crops; "harries" waterbirds and other wetland birds by flying a metre or so over grass to alarm prey; takes ducklings, frogs, small reptiles and, at dawn and dusk, rats and mice

nesting: bends over and tramples down reeds of swamp, usually well out from shore in water up to 1.5 m deep; basis of a wide platform nest onto which the eggs are laid

abundance: common

Australasian Grebe *Tachybaptus novaehollandiae*

name: Australasian Grebe *Tachybaptus novaehollandiae*
length: 23–25 cm
looks: small grebe; dark back; in breeding season black head with yellow oval on face; red down neck; non-breeding dark brown on crown
voice: fast, loud, quite high, metallic chittering; mostly in breeding season
habitat: most freshwater wetlands from large lakes and estuaries to small swamps and ponds; hunts small freshwater yabbies and other small aquatic prey
nesting: floating mass of water weeds, attached to reeds or branch reaching into water; eggs covered by departing bird
abundance: common

This is one of the most common small waterbirds and is often seen on lakes, swamps, dams and the ornamental lakes of suburban parks. It is easily recognised being significantly smaller than any of the ducks. In its breeding plumage the Australasian Grebe is easily identified by the dark head and neck, with a chestnut streak down the neck, and yellow patch around the base of the bill.

These grebes float high on the water, often with fluffed out back feathers to the sun in cool weather. Mostly they are busy little birds, spending the day plunging beak-first to the bottom of their lake or pool, 3 m or deeper, to capture yabbies or similar small freshwater crayfish. These are brought to the surface, and thrashed about with violent shaking, before being swallowed whole.

The Australasian Grebes are far better swimmers, on the surface and underwater, than almost any other waterbird, effortlessly vanishing beneath the surface of the water to reappear far from any danger.

The nest of this grebe (left), similar to that of the Hoary-headed Grebe, is a collection of water weeds piled onto any support or anchorage, such as the tip of an overhanging branch that dips into the water, a submerged log that just breaks the surface, or a small clump of reed or water lilies well out from the shoreline.

Both sexes make seemingly endless trips carrying small pieces of plant debris to pile up around this support, resulting in a floating mass which, being waterlogged, is mostly submerged with a slight hollow on top, barely above the water. The four to six eggs, initially bluish-white, soon become stained blotchy brown.

Whenever the incubating bird is about to leave the nest, it reaches over the side and drags up wet plant debris from the waterline to cover and completely conceal the eggs. If there is some danger, this urgent covering takes but a couple of seconds, and the grebe then dives overboard leaving a nest that appears empty.

The tiny striped chicks can swim and dive within a very short time of hatching. The first hatchlings swim near the edge of the nest until all the eggs have hatched. The whole family then begins to explore the lake with their parents. Often the tiny chicks will climb onto the backs of the adults to rest or hide under their folded wings.

Great Crested Grebe *Podiceps cristatus*

This large and distinctive grebe has a wide distribution through Europe, Asia and Africa and is famed for its elaborate and graceful courtship displays, which include the "weed-dance" and "penguin dance".

Great Crested Grebes are usually silent, but may very occasionally give a harsh barking threat call and some growling, crooning and grunting sounds during their displays.

The breeding season is variable, occurring sometime between August and March. The nest is a floating mass of water weeds and other vegetation, attached to either a drooping branch that touches the water or to standing reeds or other anchorage that stops the nest from drifting. Typically it is in water at a depth of 1–2 m, and is anchored 2–10 m out from the shore line.

The top of this vegetation mass is only some 10–15 cm above water level, shaped as a shallow depression. The three to five eggs are covered with water weeds when the incubating bird leaves the nest.

name: Great Crested Grebe *Podiceps cristatus*
length: 47–61 cm
looks: could not be mistaken for either of two small grebes; only species with such long neck and crests around the head
voice: usually silent; calls when breeding or if nest threatened
habitat: large lakes, river estuaries, sheltered bays; rarely on small water bodies; dives underwater to forage; prey includes fish, insects and their larvae, crustaceans, tadpoles and other aquatic life
nesting: large platform of semi-floating, flattened reeds, piled with water weeds
abundance: uncommon and only in locations water is suitable

Hoary-headed Grebe *Poliocephalus poliocephalus*

When in breeding, plumage on this small grebe's head is heavily streaked, giving it the distinctive "hoary" appearance for which it is named.

Hoary-headed Grebes are much more communal than Australia's other two grebe species. They often gather together in large floating rafts on the water and nest in groups (rather than isolated pairs). They prefer wide expanses of open waters and, unlike the Australasian Grebe, will often gather on coastal estuaries and marine waters. Rarely will any be seen on small water bodies, such as farm dams, which often support a single pair of Australasian Grebes.

This species is highly nomadic, travelling quite long distances to reach other waters. During the breeding season their groups are smaller and more scattered, but by autumn they tend to congregate on the larger lakes and estuaries.

The nest is a floating mass of water plants, usually further out from the shore than the nests of other grebes. Reeds or other plants are used to attach the nest and keep it from drifting.

name: Hoary-headed Grebe *Poliocephalus poliocephalus*
length: 29–30 cm
looks: streaked head in breeding season; dark cap down to bottom of eye; dark streak right down back of neck; back light brown
voice: usually silent, except soft sounds at nest
habitat: wetlands, both freshwater and marine (including mangroves); gathers in large flocks on water as rafts of birds; feeds on yabbies and other crustaceans, water snails, dragonflies
nesting: mass of water plants; attached by reeds to keep from drifting
abundance: common, nomadic

Comb-crested Jacana *Metopidius gallinacea*

name: Comb-crested Jacana *Metopidius gallinacea*

length: 20–27 cm

looks: unmistakable; unique in Australia for having enormously long toes, able to spread bird's weight over sufficiently large area of leaves so it can walk across lily leaves without sinking

voice: various piping and twittering sounds; a sharp, loud alarm call

habitat: lakes, swamps and dams with some substantial covering of floating water vegetation, typically large water lily leaves; captures small insects and similar creatures from the surface of lily leaves, reed leaves and from water surface

nesting: eggs laid on floating leaves within a small circle of thin, soft plant stems to stop eggs rolling off

abundance: locally abundant, especially where there are large permanent wetlands

This bird is superbly adapted to its habitat of well-vegetated wetlands, permanent lagoons, dams and other waterways where large leaves of lilies and other floating vegetation cover much of the water. Extremely long toes enable these birds to spread their weight widely for sufficient support, so that they can move rapidly, running across lily leaves, stopping, turning and probing for small creatures above or just beneath the surface.

The very long legs do not slow these birds down as they move across the water vegetation, for they can be agile and fast. Taking off at flight however, they seem initially awkward, as if held back by the drag of trailing legs and toes. With wings beating rapidly some speed is built up, and the flight soon becomes much more relaxed and graceful.

Jacanas may be seen alone, in pairs or groups, and are often very active, with much calling and chasing. They become quite aggressive towards each other in the breeding season, with each pair defending their territory.

The nest is a collection of floating aquatic plants, most of the mass below the water and held in position by surrounding vegetation. The slight depression made for the eggs is barely above water level. With the added weight of eggs and a brooding adult, the eggs can be damp or slightly submerged, although the surface water should be warm under the late spring or summer sun.

The three or four eggs are a glossy olive-tan, with random squiggly lines and streakings of deep chestnut that imitate the appearance of the surrounding expanse of floating plants. Eggs are usually only found by watching where the female sits to brood; however, she is usually wary about giving away her position when anyone is in sight, so it can be a long wait. The chicks are hatched with legs and toes that seem even more disproportionately long than those of the adults. Upon hearing the alarm call of other jacanas in the vicinity, both chicks and adults are quick to dive: there they can peek out from under leaves, but risk being taken by a water python if they remain submerged too long. Adults soon move both eggs and tiny, newly hatched chicks to a new nest, perhaps a necessary move when predators have been very close, or the water level is changing.

Azure Kingfisher *Alcedo azurea*

The tiny Azure Kingfisher is hard to spot in its shadowy haunts where trees overhang banks of creeks and rivers. Probably there are many people who do not know that their backyards, or some place nearby, is home to this small but spectacular looking bird.

Where streams flow through any of the larger properties of Brisbane's outer suburbs, a series of high, squeaky "peee-peee" calls are often the first indication of the presence of this small kingfisher. For the great majority of Brisbane residents who do not have a stream in their backyard, there are many rivers and creeks in parklands where, with patience, this species might be seen.

The Azure Kingfisher will rarely be found along waterways that have been denuded of vegetation, or have muddied waters; it prefers well vegetated banks, overhanging screening branches and logs in the water. Its favoured hunting sites will have clear water, either flowing or still, where, from an overhanging perch, this sharp-eyed bird will be able see the small fish, yabbies and other aquatic prey that will be beneath the surface.

Much of the time this kingfisher sits patiently on a perch that is usually quite low over the water, below the level of the rim of the creek banks, and often screened by overhanging or drooping foliage. There it waits patiently, motionless but for an occasional movement of its head, and almost incessant regular flicks of its stumpy tail. Its rich deep blue colouring is lost in the shadows and even its sudden departure, skimming low over the water, may not be noticed.

A sharp squeak of sound is mostly given in flight. If an open stretch of water can be watched, the kingfisher may be seen darting upstream, or down, skimming close above the surface, almost never out over dry land. Occasionally, a perched bird screened from view may plunge headlong into the water, and perhaps emerge with a fish or other prey item in its bill.

The nest is a small tunnel built into a creek bank, into which the kingfishers dart, approaching at speed, braking with broad, out-held wings, which are folded at the last moment as the birds dart into the small tunnel opening.

name: Azure Kingfisher *Alcedo azurea*

length: 17–19 cm

looks: tiny kingfisher; upperparts deep blue; neck to undertail deep rufous-buff

voice: usually silent for long periods; occasional extremely high, long squeak; in breeding season rapid series of short sharp squeaks

habitat: creeks, swamps, lakes and mangroves where banks are densely vegetated; dives from perch to take yabbies and similar crustaceans, small fish

nesting: narrow tunnel dug into creek bank leading to chamber where the eggs are laid

abundance: common where the habitat remains suitable for hunting and nesting, otherwise uncommon

Chestnut-breasted Mannikin *Lonchura castaneothorax*

name: Chestnut-breasted Mannikin *Lonchura castaneothorax*
length: 11–12 cm
looks: typical finch shape; unique black face; orange-chestnut breast; black breast band; white belly
voice: contact call a high "tlit"; also has a series of high clear notes
habitat: damp environs, swamp edges, reed beds, grassy woodlands; climbs grass stems to strip away seeds as they begin to ripen
nesting: small domed nest in reed bed clump
abundance: common; nomadic (so variable seasonally)

Outside its breeding season the Chestnut-breasted Mannikin is a highly sociable species, forming large nomadic flocks that bunch together and change their flight direction in unison. To do this they cease their usual undulating finch flight and travel fast and direct, which enables each bird to synchronise changes in direction, the flock twisting and turning as one.

These birds not only travel together, but keep in flocks to feed, to visit water, to drink and to sleep. Unlike some other finches, however, they do not build roost nests, but cluster together within dense clumps of reeds.

So strong is their flocking that pair bonds seem weak by comparison; there seems to be little recognisable interaction between partners. The species also remains very sociable throughout the breeding season. Rather than stay in their very large flocks, they separate out into smaller, but still close-knit local breeding colonies, nests closely clustered, often only a metre or so apart.

In their flocks the Mannikins give a clear bell-like "tlit" — a longer "tleit" is used in feeding flocks. These birds also have a song — a series of fine, clear notes delivered in many variations.

This bird inhabits areas of tall rank grasses or reeds around the margins of swamps, rivers, creeks, overgrown drains, wastelands, coastal heaths, lantana thickets, cereal crops, pastures and mangroves. It feeds on the seeds of grasses, stripping them from the heads before they are fully ripened. Insects are caught in flight, the Chestnut-breasted Mannikin relishing flying termites.

The nest of this species is quite small and domed. It is constructed from dried grass placed within a dense clump of tall grass, reeds or within a bush. The nest has a downwards-sloping entry tunnel at one end. The four to six eggs are white.

Unlike many birds of the grass-finch family, Chestnut-breasted Mannikins do not build roost nests in the non-breeding season; rather, large flocks crowd together in dense tall grass or reeds. During the breeding season however each pair or family group will roost in breeding nests. Once all young are flying, nests are abandoned in favour of the communal roost sites.

Buff-banded Rail *Gallirallus philippensis*

This is a beautifully coloured and patterned bird found in the swamps and reedy margins of lakes, where it is seldom seen by the casual observer. Like those other smaller and equally secretive inhabitants of dense wetland vegetation, the crakes, the Buff-banded Rail presents just a little more of a birdwatching challenge.

Rails and crakes are out and about at first light, when they emerge from their dense swamp vegetation to forage on open mudflats between the reed beds and the shore. In the pre-dawn dim light rails and crakes wander where the shallows of the swamp or lake are for them little more than knee-deep — perhaps 2 cm for the crakes, and slightly deeper for the larger Buff-banded Rail. But always, they are ready at the slightest disturbance to dart back to the safety of their hideouts.

So the challenge is to be there very early, waiting as the eastern sky begins to lighten, with an intention to stay until the sun is fully over the reeds and mudflats. The late afternoon or early evening also provides an opportunity to see some of these secretive birds out foraging.

However at some sites, local parks or other reserves where people commonly visit, or picnic close to water and reedbeds, these birds sometimes become less timid. They are more inclined to venture out in the open on nearby mudflats or lawns.

The Buff-banded Rail has beautifully coloured and patterned plumage. The adults sport a rich, bright chestnut band through the eye and down each side of the neck and a patch of chestnut across the breast. Immature rails have comparatively dull plumage with buff bars and they lack the colourful chestnut so prominent in the face and breast of adults.

Most remarkable are the chicks, which hatch clad entirely in black down and leave the nest soon after hatching. Following in their colourful parents' footsteps, waiting to be fed, the rail families present an incongruous picture with their vastly different plumages.

The Buff-banded Rail hides its nest in a dense clump of grass or other vegetation, sometimes quite distant from the water. The nest is a deep bowl of reeds or grass, tucked into a slight hollow of the ground. The clutch of five to eight eggs is incubated by both adults.

name: Buff-banded Rail *Gallirallus philippensis*
length: 28–32 cm
looks: very colourful, tail-standing rail; upperparts chestnut; orange and white streaks across face; underparts heavily cross-barred except bright chestnut patch on breast; chin to breast plain blue-grey
voice: usually silent, but other sounds given in swamp reed beds
habitat: dense, damp vegetation around swamps, lakes, creeks, coastal lagoons; seeks small items of animal life on mudflats, wet grass, shallow water
nesting: bowl of grass or reed leaves hidden in dense vegetation
abundance: quite common, though shy and not often seen

Yellow-billed Spoonbill *Platalea flavipes*

name: Yellow-billed Spoonbill *Platalea flavipes*
length: 75–90 cm
looks: unmistakable; long, pale yellow spoon-tipped bill
voice: usually silent; in threat displays gives soft coughs or grunts
habitat: shallow swamps, fresh or brackish water, flooded pastures, small pools and dams (fresh or brackish); uses spoon-tipped bill to find and capture small edible creatures from muddied shallows
nesting: in colonies; nests in trees with other spoonbills, egret and herons
abundance: common in near-coastal wetland areas, and inland after heavy rain

The Yellow-billed Spoonbill, like the Royal Spoonbill, feeds by swinging its long, flat-tipped bill slowly from side to side through shallow water. The sensitive bill tip is able to detect small prey in the most muddied waters. Differences in the feeding habits of Yellow-billed and Royal Spoonbills allow each to exploit a slightly different food source and create less competition between the two species. Yellow-billed Spoonbills spread out to many small freshwater wetlands, feeding mainly on water insects in shallow water day and night. They also eat some small fish, crustaceans and molluscs. Royal Spoonbills feed in groups or flocks on larger areas of water, including tidal mudflats.

These birds return to the same breeding area each year, (provided there is sufficient water), nesting as solitary pairs or in colonies in freshwater wetlands. The nest (a wide shallow platform of sticks lined with finer twigs and paperbark) is built in trees or shrubs standing in water or on trampled clumps of reeds.

Royal Spoonbill *Platalea regia*

name: Royal Spoonbill *Platalea regia*
length: 75–80 cm
looks: unmistakable; long, black spoon-tipped bill
voice: silent, except for some soft, low grunts at nest
habitat: shallow swamps, fresh or brackish waters, floodwaters; rarely tidal waters; sweeps water with slightly open, sensitive bill-tip for tiny food items and aquatic life normally hidden in muddied water and soft mud
nesting: usually in colonies; large stick nest in tree standing in water
abundance: quite common in suitable habitat

Royal indeed — a tall and stately bird, very formal, with flowing white plumes behind the head when breeding. It is a gregarious species, flying in small flocks, often in lines or in a "V" formation. Usually seen feeding in small groups (perhaps five to ten), working their way through the shallows. Royal Spoonbills use shallow wetlands and the margins of deeper wetlands (fresh, saline, open or densely vegetated waters). They are also found around coastal lagoons and mangroves. This is a common species in most parts of its range.

Spoonbills are silent except at the nest site when breeding, where they give off soft, low grunts, groans and hisses. Breeding occurs October to March in the Greater Brisbane region, usually in colonies of ten to fifty birds, loosely congregated and intermixed with other nesting waterbirds. The nest is a large and bulky dished platform of sticks and reeds built into a tree standing in water or near water, or it is built low over the water on reed beds.

Eurasian Coot *Fulica atra*

This species is one of the most common waterbirds on lakes, reservoirs, rivers, dams, and on ornamental pools in suburban Brisbane. It also uses temporary floodwaters, farm dams and irrigation storages, and only rarely appears on marine wetlands.

The Eurasian Coot is easily identified by its rotund, blackish body and the unique white shield that continues unbroken from the bill high up onto its forehead. Its red eyes look all the more brilliant surrounded by its black face. Calls include an abrupt loud "krek", a sharper "krik", and rapid, sharp grating "kiek-kiek-kiek".

The Eurasian Coot is a highly aquatic species, well adapted to life on and under the water. Its feet are set far to the back of the body (like boat's propellers) and are ideal for propulsion, especially underwater. The Eurasian Coot is a vegetarian, upending or diving to bring up edible water plants.

name: Eurasian Coot *Fulica atra*
length: 35–38 cm
looks: blackish rather plump-bodied waterbird with white bill and forehead shield, bright red eye
voice: mostly an abrupt sharp "krek" or "krik"; grating "kierk-kierk"
habitat: most freshwater wetlands, occasionally saline estuaries. Feeds on aquatic plants by upending, or diving where water is deeper.
nesting: semi-floating nest in shallows, built up of leaves of reeds and rushes, or floating, in which case anchored to a branch or other support
abundance: very common

Purple Swamphen *Porphyrio porphyrio*

The Purple Swamphen is a large, colourful bird that dominates swamps and surrounding wet grasslands. The stout red bill and forehead shield, coupled with its red eyes, form a bold colour contrast against the blue-black of the bird's head, neck, and the deep indigo of its breast.

Unlike most waterbirds, the Swamphen does not have webbed feet, so it swims quite slowly, and never below the surface. It can run fast, and in flight it is strong if clumsy. With its speed on land and a powerful bill, the Purple Swamphen is a threat to many smaller birds of lakes and swamps. But in the water it can catch very few prey, except by surprise. Ducks, grebes, and the Eurasian Coot all easily outswim the Purple Swamphen, or they can readily escape predation by swimming underwater. The Purple Swamphen has a variety of booming and shrieking calls, mostly at night. It also makes a thudding noise by clapping its wings against the body. The nest is a semi-floating mass of vegetation, heaped upon bent-over reeds, which will typically contain three to eight eggs.

name: Purple Swamphen *Porphyrio porphyrio*
length: 45–50 cm
looks: large hen-type bird; head and underparts deep purplish-blue; massive bill; forehead shield and eyes bright red; legs and toes long; feet not webbed
voice: great variety of noises; abrupt "hak"; harsh, sharp "kiaark"; noisiest at night with wild shrieking and booming sounds
habitat: most wetlands, freshwater and sheltered tidal areas, including similar habitats in parks and gardens
nesting: builds in a clump of reeds, bending reeds over, adding vegetation; in shallow water pile of vegetation built up from bottom of water
abundance: common

Heathland Birds

Most heathlands support a rich and varied bird population. A wealth of plant species, nectar-bearing flowers and dense cover together provide a very favourable habitat, especially for honeyeaters, wrens and other small passerines.

Heaths are communities of shrubs of many families; trees and shrubs found there are usually less than 2 m in height and often under 0.5 m. Foliage tends to be hard and often prickly. Grasses and herbaceous plants are rare. Coastal sand heaths are the most widespread examples of this habitat. Of these, the massive Fraser Island sand dunes north of the Greater Brisbane region are the best known, but Great Sandy National Park, Bribie, Moreton, North and South Stradbroke Islands, and parts of the mainland also feature this type of heath.

Along this coast, sandy soils inland from the beaches have evolved a type of vegetation commonly known as "wallum". Here the ground is densely covered with low scrubby plants exceptionally rich in wildflowers, with trees that tend to be small and sparsely scattered. Density of shrub cover varies greatly, allowing descriptions of heaths as closed or open. Often trees, banksias, tea-trees, eucalypts or casuarinas, are scattered through the heathland, and where these trees become more numerous heathland can blend to woodland with the heath continuing as the understorey.

Within the broad heathland habitat are several distinct types. These include sandplain-heaths on low fertility sands, both coastal and inland, in localities ranging from high to low rainfall, and swamp-heaths which include alpine and button-grass types. These swamp-heaths, waterlogged for part of the year, tend to have a poorer variety of bird species, while the sandplain-heaths have a larger list, especially where they intermix with open banksia, mallee banksia or mallee, and share many of their species.

Heathlands tend to have fewer bird species than forest habitats; there appears to be a correlation between the number of layers of vegetation and the total species that can be supported. Forest, with at least two and often three strata, can support many more species than the single stratum of heathland; the more complex the vegetation structure, the richer its avifauna tends to be.

Heathlands are not limited to coastal or sandy soils and can occur in layerings of forest or woodland, sometimes where there is very shallow soil preventing growth of trees or tall shrubs: the only ground cover is a dense layer of small plants collectively known as heath. These may consist of entirely different plant populations than those associated with the heaths of coastal wallum country.

TOP BIRDWATCHING sites

COOLOOLA — GREAT SANDY NATIONAL PARK

This is an extensive tract of sandy country — some 35 km north–south by 25 km east–west. It has a rich and diverse environment, including a large extent of wallum heathlands, shallow sandy paperbark swamps and areas of rainforest growing on sandy soil. With such a diverse habitat the area is exceptionally rich in wildflowers, which attract a great variety of nectar-eating birds.

BRIBIE ISLAND NATIONAL PARK

Set close to the mainland at the northern end of Moreton Bay, Bribie Island has about one-third of its 4700 ha reserved as national park. Birds are abundant in this banksia and heathland country, especially when the many heathland plants are in flower through the spring months.

NOOSA NATIONAL PARK

A small but vital national park that preserves some of the rich and varied natural environment of a coast rapidly being taken over by coastal development. The wallum areas have a rich and diverse flora, and many wildflowers bloom in spring. For this reason, spring is the best time to birdwatch.

THE NOOSA BIRD TRAIL

A bird trail additional to the national park, with stops at 32 sites in varying habitats where a substantial list of bird sightings might be made.

Right: Variegated Fairy-wren.

Yellow-tailed Black-Cockatoo *Calyptorhynchus funereus*

name: Yellow-tailed Black-Cockatoo *Calyptorhynchus funereus*
length: 58–65 cm
looks: large; all-black cockatoo; no other black-cockatoo with yellow ear-patch and yellow in tail
voice: far-carrying "why-eela, weee-la"; harsh alarm screeches
habitat: diverse habitats, coastal and inland; heath woodlands, eucalypt forests and woodlands, rainforests; searches for seeds of native vegetation; pine plantains for seed and wood-boring grubs
nesting: very high, large and deep hollow of forest tree
abundance: moderately common, locally nomadic, or migratory

This bird is usually noticed when a flock passes overhead. Travelling with a typically languid, undulating flight of deep lazy wingbeats and floating glides, flocks of these birds weave uncertain pathways among, around and over the treetops. Meanwhile, their wailing calls carry far and draw instant attention to the passage of the flock overhead. The call of the Yellow-tailed Black-Cockatoo is a far-carrying "why-eee-la, weee-la". When alarmed, it responds with loud harsh screeches.

This species of black-cockatoo appears to be quite common. It may still be seen over Brisbane suburbs when local trees carry the seeds that attract these birds to the city's forests, parks or gardens. Often such flocks are moving overhead on nomadic wanderings that take them into many different habitats.

Many of the native tree and shrub species that the Yellow-tailed Black-Cockatoo visits when their seeds are ripe include species of banksia, hakea, grevillea, acacia and other trees and shrubs of coastal heathlands. At other times these cockatoos move to eucalypt forests, where seeds of eucalypts form the focus of their feeding.

Pine plantations have become an important food source for these birds, and Yellow-tailed Black-Cockatoos are often found among those trees, feeding not only on their seeds but also on wood-boring grubs. The cockatoos reach them by tearing away the surface bark with their powerful bills to expose the juicy, wood-borers in their tunnels.

For its nest, this species uses a high, massive limb or large hollow of a tree trunk, where the floor of the hollow measures up to 40 cm across. Two eggs are laid, several days apart, but the female begins incubating immediately after laying the first. Upon hatching, the second chick is several days younger and significantly smaller and weaker than its sibling. It usually fails to compete for food, and dies within a day or two.

The surviving chick is brooded by the female for a month, as the male comes to the nest to feed her. Once the chick is large enough to be left alone, both adults bring food and the chick finally leaves the nest hollow at about eleven weeks of age.

Horsfield's Bronze-Cuckoo *Chrysococcyx basalis*

Australia's bronze-cuckoos are named for the shiny appearance of much of their plumage. Over the underlying brown tone there is a surface gloss of iridescent green, most evident on the wings, with the combined brown and green giving this species its bronze appearance.

The brownish-bronze colour forms bars across most of the bird's white underparts from its throat to undertail. These bars are broken down the centre line, displaying a continuous white strip running down the centre of the breast, belly and undertail. This is one of several features by which this species may be separated from the similar looking Shining Bronze-Cuckoo.

Horsfield's Bronze-Cuckoo also has a blackish line through its eye back down to its neck; the Shining Bronze-Cuckoo has only a vaguely darker patch behind its eye. Also important in identification (when visible) are the rufous edgings on the outer tail feathers. In contrast, the Shining Bronze-Cuckoo sports a greenish-brown upper tail and a white-barred, grey-brown undertail.

These two similar bronze-cuckoos can also be separated by calls. The Horsfield's Bronze-Cuckoo gives long, slow descending whistles ("tsieeeuw, tsieeeuw"), while the similar-looking Shining Bronze-Cuckoo has a series of short quick whistles ("hwee, hwee, hwee"), each rising quickly from low to high pitch, these followed by a longer wavering downward "pheee-ier" rather like the downward call of the Horsfield's.

This cuckoo uses a great variety of habitats, including both low scrubby heath and patches of heath growing as ground cover across sandplain woodlands. It may also be encountered in forests, mallee, and semiarid inland habitats. In urban areas it may be seen in parks, gardens or roadside vegetation.

The Horsfield's Bronze-Cuckoo lays its eggs in the domed nests of various small birds, including fairy-wrens and thornbills, and also in the open cup nests of species that lay speckled eggs similar to the eggs of this cuckoo.

The host species include various honeyeaters, which are common across heathlands during spring when banksias, bottlebrushes and grevilleas are in flower.

name: Horsfield's Bronze-Cuckoo *Chrysococcyx basalis*
length: 14–17 cm
looks: upperparts brown with green iridescence; breast to undertail white with bars; bars have central gap right down centre line; dark streak through eye; rufous edge to outer tail feathers
voice: piercing whistle; descending "tsieeew, tsieeew"
habitat: heathlands, woodlands, open forests, farmlands, gardens; hunts from perch, dropping to ground to take insects
nesting: parasitic; lays eggs into domed nests of small birds
abundance: common; migrant to southern Australia

Common Bronzewing *Phaps chalcoptera*

name: Common Bronzewing *Phaps chalcoptera*
length: 30–36 cm
looks: large-bodied, small-headed pigeon; sides of neck blue-grey; colourful green and bronze iridescent speculum panel in folded wing
voice: slow, mournful, deep, resonant "whooo", repeated many times; in display a more energetic, abrupt "whoo, hoo-hoo"
habitat: widespread in most habitats; forests, woodlands, farmlands, outer suburbs, parklands; usually in small groups; peck about on ground for seeds
nesting: flimsy, rough stick nest on a horizontal perch
abundance: usually common, locally nomadic

The Common Bronzewing is the pigeon most associated with heathlands and other more open habitats, such as dry eucalypt forests, especially where there is heath undergrowth in woodland or forest. This bird is rare in dense wet forests.

Bronzewings are ground-foraging pigeons with a plump-bodied, small-headed appearance. Whether in small groups, in pairs, or alone, they peck about on the ground beneath heathland shrubbery, finding seeds dropped by native shrubs. These birds are extremely widespread (except for rainforests), existing in most dry land habitats.

Bronzewings are extremely wary birds. At the first sight of possible danger (often quite distant) they will freeze — their camouflaged plumage making them difficult to see. If approached closely, or met unexpectedly close, they take off with a startling action — an almost explosive steep takeoff with a loud clatter of wings. Often the presence of one or more of these pigeons, out of sight beneath the undergrowth, will be marked by a "whooo" or "ooom" repeated monotonously for long periods of time.

When observed in flight, it is most likely to be a single bird or several birds in fast, direct movement travelling low across heathland or else dashing away among woodland trees. Their swift passage is made obvious by the whistle of their wings — their identity confirmed by the cinnamon tone of their underwings displayed momentarily at the point of uplift before each powerful downbeat. To warm themselves, bronzewings often sunbathe with their wings outspread (see photo at left).

Bronzewings come to water each morning and evening. With typical wariness, they will often alight in a tall tree overlooking the water, watch for lurking danger, then flutter down to begin their slow, cautious walk over the last 30–100 m to the water's edge. Sensing any danger, they will take flight explosively with a loud clapping of wings. Of all the builders of stick nests in trees the Common Bronzewing must be one of the worst, making a small, crude unlined shallow saucer of thin sticks. The floor of the nest may be so thin that the eggs can be seen through its sparse open mesh; other nests, however, may be quite substantial.

Red-backed Fairy-wren *Malurus melanocephalus*

This is the smallest fairy-wren, but no other species of Australian fairy-wren has this bold combination of scarlet and black. With the arrival of spring, breeding male Red-backed Fairy-wrens appear cocky — seemingly assured by the power of a colour combination so often displayed in nature as a warning worth heeding. These males dart about, often calling, and displaying aggressively towards each other.

For all their bluster, males quickly reach some sort of truce, compromise, or mutual respect for territorial borders, and settle down to the responsibilities of raising a family. The females are soon busily engaged in nest building. The nest is usually placed close to the ground, often in tall grass, or in dense low shrub.

Fully coloured males are greatly outnumbered by brownish birds — immatures, females and males in eclipse plumage. In the autumn the males moult to eclipse plumage and become brown, similar to the female. Older dominant males moult direct to full colour, breeding plumage, bypassing the eclipse stage.

Red-backed Fairy-wrens can come visiting or take up residence in the smallest of inner-suburb backyards. This is much more likely if there is any adjoining or nearby habitat suitable to make a territory large enough for a family of fairy-wrens. Fairy-wrens may be small but require a constant supply of tiny insects. Further out from the city, larger backyards are even more likely to support these birds. As with the small backyard, the larger block will be more appealing to fairy-wrens if it is close to a forest, parkland, grassy roadside, creek, or other blocks of land that can serve as additional habitat.

The preferred habitats for Red-backed Fairy-wrens include woodland and open forest, the edges of rainforest, tall grass among shrubs and trees, and areas of low scrubby thickets often of lantana.

People fortunate enough to have these delightful birds visiting or living near their homes have backyard company well worth encouraging. Like other birds of low-level vegetation, these wrens will probably most often use parts of a backyard that have a mixture of tall grass and low shrubs — perfect habitat for a Red-backed Fairy-wren family.

name: Red-backed Fairy-wren *Malurus melanocephalus*
length: 10–13 cm
looks: male unique; entirely black except for brilliant red back; female plainest of all female fairy-wrens; light sandy-brown above; sandy-buff beneath; tail plain brown; face lacks markings
voice: very high weak squeaks, insect-like, followed by a stronger (but still not very loud) trill
habitat: heathlands, woodlands, open forests, parks and gardens
nesting: builds a small domed nest of grass and fine strips of bark; built very low in a clump of grass or low shrub
abundance: common and widespread

Variegated Fairy-wren *Malurus lamberti*

name: Variegated Fairy-wren *Malurus lamberti*

length: 13–14 cm

looks: breast of male is black (not dark blue); chestnut-shouldered; violet-blue crown; female has pale brown upperparts; dark chestnut around eyes; tail has only slight tint of blue

voice: abrupt "trrt-trrt-trrt" contact call; song, only in breeding season, is rapid, mechanical, metallic trill at fairly even pitch

habitat: wide-ranging species; uses a great variety of habitats including coastal heaths, undergrowth of forests and woodlands; forages among foliage and on ground for insects and small spiders

nesting: small domed nest, with side entrance; low in undergrowth or in tall grass clump

abundance: moderately common, usually sedentary

The Brisbane region is home to three species of fairy-wren. Unlike fairy-wrens in some regions of Australia, where several species may exist and are hard to tell apart, each of the Greater Brisbane region's fairy-wrens is distinctive and easily identified.

Fairy-wrens are instantly recognisable — the long slender tail, carried vertically, is an obvious clue to their identity. Beyond that, the colours and patterns differ greatly. All are jaunty little birds that, with tails held high over their backs, bounce across the ground rather than run, and fly low with long tails streaming behind. While the fully plumed males of Variegated, Red-backed and Superb Fairy-wren species are easily separated, the females and non-breeding males are not so easily identified.

Of all the fairy-wrens, the Variegated has the widest distribution, covering most of the Australian continent. The great diversity of vegetation and climate across this vast area has caused slight variations that have been named as subspecies. The Variegated Fairy-wrens of south-eastern Queensland were the first named and are therefore known as the "nominate" race.

Variegated Fairy-wrens are not migratory or inclined to wander far, but keep fairly close to their territory most of the year. Any backyard fortunate enough to support a family of this species would almost always have them around, or else not too far away. However, for these and any other birds that forage on, or near the ground, and nest low, a predator-safe garden is vital. For example, a cat on the loose during daylight hours can wreak havoc on these birds. At night, these fairy-wrens tend to roost at a safer distance on higher vegetation.

Often the presence of these birds is revealed by their sharp, high calls. The voice of the Variegated Fairy-wren has a squeaky and rather metallic quality; most of the year it is an abrupt "trrt-trrt", but a more prolonged song sounds in the breeding season — a rather mechanical rattling trill "trrrt-trrrt, trrrt-trrrt". Short, sharp contact calls are made while foraging.

The breeding season extends from August to February. The nest is a small, rounded, domed-roof structure, built low in a clump of coarse grass, or in a low dense shrub.

Yellow-faced Honeyeater *Lichenostomus chrysops*

This species is a busy, active honeyeater found in forests and woodlands, heaths, and mangroves. One of Australia's most migratory species, thousands of Yellow-faced Honeyeaters leave their breeding grounds in autumn, moving north between March and May in flocks of hundreds or, in some instances, more than a thousand birds. (Often these flocks include other honeyeater species, pardalotes and other small birds.) However, not all honeyeaters migrate; some birds remain in their region, so that in all parts of the habitual territory there will be a smaller resident population during the migratory part of the year.

At the northernmost parts of their range, flocks of Yellow-faced Honeyeaters wander in search of nectar and manna. They glean insects from foliage, and also take insects from bark, from flowers, and in flight. In some localities, manna from the manna gum is taken. Their reach includes most kinds of habitat, whether forests, woodlands or heathlands of the high ranges or coastal lowlands.

Yellow-faced Honeyeaters begin returning to their breeding grounds in mid-eastern and southern forests and woodlands from July. This return is less conspicuous, the birds travelling in smaller flocks, moving more slowly, until they reach their previous breeding territories.

Yellow-faced Honeyeaters mate as permanent pairs that together defend a nest territory. The nest may be built in the foliage of an understorey shrub, in the outer foliage of trees, and in garden trees and shrubs. The female gathers nest material, but is often accompanied by the male. The nest is a small cup-shaped structure of grass and bark fibre well bound with webs and softly lined. Its rim is attached to foliage twigs. The female alone incubates, but both sexes feed the nestlings.

The song of the Yellow-faced Honeyeater is commonly described as spirited, melodious, and cheerful. Included is a cheery ringing "chwikup, chwikup", and a rollicking song, "whit-chiwit-chiwikup". In flight the Yellow-faced Honeyeater uses a soft "klip-klip" to help keep the flock together, while at other times it uses a rather peevish "kreee" to maintain contact.

name: Yellow-faced Honeyeater *Lichenostomus chrysops*

length: 16–18 cm

looks: small; overall grey-brown; paler beneath; bright yellow streak from bill to eye then down over ear and set between bold black streaks back from brow and chin

voice: cheery, ringing "whit-whit" and "chwikup"; brisk, rollicking song "whit-chiwit-chiwikup"

habitat: heathlands, woodlands and forests from coastal mangroves to high ranges; forages in trees for insects of flowers, foliage and bark; nectar and manna eaten in summer and autumn

nesting: small cup nest; suspended by rim from twigs among foliage of tree or shrub

abundance: common, migratory; large flocks move north in autumn to spend winter in SE Qld and further north; return to southerly forests in spring

White-cheeked Honeyeater *Phylidonyris nigra*

name: White-cheeked Honeyeater *Phylidonyris nigra*

length: 16–19 cm

looks: medium sized honeyeater; conspicuous yellow panel in folded wing; white streaks at brow; very large, pure white cheek patch; similar to New Holland but two, not three, white marks each side of head; thicker bill

voice: quite unlike New Holland; distinctive yapping, not usually high pitched "ch-quickup"

habitat: heath-like or shrubby undergrowth of eucalypt forests and woodlands; heathlands

nesting: small, sturdy, softly lined cup nest; usually very low in heath in dense shrub

abundance: quite common in heath habitats, uncommon elsewhere; locally migratory

Although the White-cheeked Honeyeater shares an overall similarity with the New Holland Honeyeater, it is quite distinctive not only in plumage, but also in its behaviour and calls.

Both species are about the same size, with almost identical body plumage. The yellow panel on the folded wing is the most obvious similarity. The really conspicuous difference between the two species is the pattern of white markings on the black head.

Both have a white eyebrow line, but where the New Holland Honeyeater has small white patches at its cheek and ear, the White-cheeked Honeyeater has a single, very large, stark-white cheek patch. Its bill is noticeably heavier — both longer and thicker.

Once their calls are known, one does not have to see either of these birds to know which is present. On the whole, the White-cheeked Honeyeater may not be as noisy, and seems less inclined to emit loud, sharp alarm calls and aggressive scoldings.

In spring, however, when many White-cheeked Honeyeaters are gathered together in flocks, they can raise quite a noise. Their calls can perhaps best be described as a rather musical yapping with two notes in quick succession, variously expressed as "chakup-chakup" or "chwikup-chwikup".

During breeding season, White-cheeked Honeyeaters perform display flights, in which the bird rises high over the nest, singing, then glides steeply down. The song at this time is a pleasant musical "chwippy-choo, chwippy-choo, twee-tee,twee-tee".

In personality, too, these honeyeaters are quite different. While the New Holland is often aggressive towards other birds, especially smaller honeyeaters, the White-cheeked has a more tolerant manner. New Holland Honeyeaters tend to stay in one locality, or make only local nomadic movements, whereas White-cheeked Honeyeaters seem to wander further in their small flocks or groups in search of nectar and insects.

The nest of the White-cheeked lies in a dense low shrub. It is similar to the New Holland's nest, but favours more bark in its construction, giving the nest thicker walls.

New Holland Honeyeater *Phylidonyris novaehollandiae*

In the early morning, tall flower spikes of banksias, bottlebrushes and grevilleas are overflowing with nectar. Many bird species dash from flower to flower, but the most conspicuous of these are New Holland Honeyeaters — noisy and often dominating, rushing threateningly at other (usually smaller) species of honeyeater looking to feed nearby.

Occasionally this fast tempo of early activity misses a beat when a loud call, sharp and urgent, rises above the steady bird chatter. In those few seconds all bird noise is silenced. Then the alarm call is repeated, from a different direction. In only a few seconds the warning has radiated out in all directions from that first call.

Moments later a raptor glides away, swift and silent, vanishing among the trees together with his chances of surprising any small, nectar-loving prey. Once spotted by that firs- New Holland Honeyeater his presence is quickly relayed by all. The predator, disarmed, has no choice but to move on. Despite a list of rowdy and rather brutish traits, the New Holland Honyeater is the sentinel of the bush — usually among the first birds to broadcast alarm at the possibility of danger.

The New Holland Honeyeater is distinguished by its black plumage, which is heavily streaked with white, and its yellow wing patch. This is a noisy species with a variety of sharp calls. Its alarm call is a high, sharp, and loud "chwiep-chwiep". When scolding its own kind or bullying o-her species, a "tjuk, tjuk" is delivered rapidly in machine-gun bursts.

The New Holland Honeyeater uses a wide variety of habitats. Forests, woodlands, heathlands, mallee and coastal thickets all form adequate homes for this species. It is also one of the most common honeyeaters in suburban gardens. It breeds mainly from spring to summer, but will happily breed at any time of year when nectar-producing flowers are in abundance.

The New Holland Honeyeater's nest is a well-made, relatively thick-walled and softly lined cup of grass and bark, built into the dense foliage of a shrub or else built rather low in a bushy tree at a height of 1–5 m. The usual clutch for this species is two or three eggs.

name: New Holland Honeyeater *Phylidonyris novaehollandiae*

length: 17–19 cm

looks: medium sized honeyeater; conspicuous yellow panel in folded wing; white streaks at brow, ear and across lower face

voice: abrupt, metallic often piercing "chwik"; alarm sequence as "chwiep-chwiep"; scolding "tjuk"

habitat: heathlands, forests and woodlands; takes nectar; insects on wing and at flower, or amongst foliage

nesting: small, sturdy, softly lined nest; usually in dense, often harsh or prickly shrub

abundance: common

Figbird *Sphecotheres viridis*

name: Figbird *Sphecotheres viridis*
length: 28–29 cm
looks: body olive-green; head black; throat and upper breast grey; bright orange-red bare skin from base of bill around eye to ear
voice: many and varied calls, commonly a loud, rising-falling, quite musical "tsee-chiew"
habitat: diverse, including rainforest margins, mangroves, woodlands, orchards, gardens
nesting: nest placed among outer foliage of a leafy tree in or near rainforest; quite small cup nest, suspended by the rim from slender twigs
abundance: common

Flocks of Figbirds, busily foraging in and about the foliage, or performing acrobatics on twigs or overhead wires, can hardly go unseen — especially adult males, so resplendent in yellow, black and red. From treetops to low garden shrubs, wherever there are fruits or berries to plunder, the Figbird can scarcely go unnoticed.

The Figbird is one of three Australian representatives (including the Olive-backed and Yellow Oriole) from a family of tropical birds that has many species distributed throughout Africa and Eurasia. The Figbird, however, is a much more sociable species, gathering in large flocks of 20–40 birds (and at times many more). These flocks converge at fruiting trees, showing a strong preference for both wild and cultivated figs.

When feeding, Figbirds clamber about in a parrot-like manner, clinging and hanging among foliage and twigs to reach ripe fruit, nectar and insects. Unlike the often solitary and more subdued orioles, a flock of Figbirds raiding a tree is most conspicuous for the constant chatter and bustling activity.

Their calls are many and varied, commonly a rising and falling, loud, quite penetrating yet rather musical "tsee-chiew", or faster "tsichew". When calls from within the flock or group are answered, many whistles soon intermingle in a confusion of sound, with individual calls seeming to be get faster and more excited. The song is softer and mellow compared with these calls, and includes some mimicry of other birds.

Figbirds use a variety of habitats, wherever suitable fruiting trees can be found. These include margins of rainforests, eucalypt forests and woodlands, mangrove and paperbark swamps, gardens and orchards.

The Figbird that visits the suburbs of Brisbane will be one of the southern race, with a grey chin and breast and its remaining underbody bright yellow. The northern race distributed across tropical Australia has bright yellow over its entire underbody. Both southern and northern races are easy to identify as Figbirds, sharing the characteristic patch of bright red skin around each eye.

Black-shouldered Kite *Elanus axillaris*

The Black-shouldered Kite almost always identifies itself within a short period of time by its superb hovering ability. It is able to maintain a position over a precise point on the ground, despite wind, while watching intently for any small movement in the grass below. It then drops almost vertically to take its prey. The only other small bird of prey with this hovering skill is the Nankeen Kestrel, which is markedly different in appearance.

When the bird is perched, the large black patch on each shoulder distinguishes it from all other species except the near-identical Letter-winged Kite which, when seen in flight from below, has a broad black band across each wing roughly forming the letter "M". (It would only rarely be seen in the Brisbane region.)

Black-shouldered Kites, while generally common, are highly nomadic, moving to districts wherever there are boom populations of mice living in grasslands or tracts of low vegetation. In the years when a prey species, usually mice, is locally abundant, these beautiful white kites can be frequently seen over farm paddocks, market gardens and grassy roadsides. Much of their rodent hunting is performed at dawn and dusk.

In the city suburbs, Black-shouldered Kites hunt over vacant lots, grassy backyards and almost any patch of grass (even those surrounded by homes or factories). When the rodent population falls, most of the kites move on and may be absent for months, even a year or so, until the mice return in large numbers.

This species breeds at almost any time of the year when its prey is most abundant, more commonly in spring to early summer months. In years of ample prey they may raise several broods. A stick nest, 40–50 cm in diameter is built into a medium-sized or tall tree — typically well hidden in dense foliage towards the top.

The female incubates the eggs, but the male often takes up guard duty perched high and conspicuous on a nearby tree. If the tree is approached, the kites may give a short, plaintive, piping "siep", and a sharper "ki-ki-ki" distress call. A serious territorial breach may cause them to swoop aggressively at the intruder.

name: Black-shouldered Kite *Elanus axillaris*

length: 35–38 cm

looks: small hawk; white and pale grey; black shoulder when perched; in flight from beneath it has dark outer wing, white inner wing; similar Letterwing Kite has black "M" shape across full length of underwing

voice: contact call "chek-chek"; faint "siep" when perched; sharp "kik-kik-kik" perhaps as distress or warning

habitat: grasslands, farmlands, vacant grassy urban lots, grassy roadsides, very low heaths; hovers over grassland and other very low vegetation; drops down and takes any small prey; mostly mice, but also very small ground birds and small lizards

nesting: nest of sticks usually in densely foliaged part of a tree

abundance: common, but depending upon prey populations may either be locally scarce or common

Scaly-breasted Lorikeet *Trichoglossus chlorolepidotus*

name: Scaly-breasted Lorikeet *Trichoglossus chlorolepidotus*
length: 22–24 cm
looks: when perched very plain leaf-green except red bill, red eye, yellowish outer tail; in flight spectacularly transformed, with most of underside of wings bright orange-scarlet and black
voice: sharp, short screeches; usually higher, finer than Rainbow Lorikeet
habitat: covers most vegetation types wherever there are forests, woodlands, paperbark woodlands, heath woodlands; seeks nectar, flowers, seeds and berries of eucalypts, grevilleas, banksias and other flora
nesting: uses small hollow of a tree
abundance: common; nomadic; abundance varies locally

Wherever trees are flowering, Scaly-breasted Lorikeets may be encountered, whether in forests of tall trees on the mountain ranges encircling the Greater Brisbane region, or on trees or shrubs of coastal heathlands. They often visit Brisbane's suburban gardens, feeding from flowering trees and shrubs.

Mixed flocks of Scaly-breasted and Rainbow Lorikeets are often seen, as both species share similar habitat and feeding preferences. They have been recorded taking nectar and pollen from trees and shrubs including grevilleas, banksias, eucalypts, coral trees and Moreton Bay chestnuts. Various seeds are also eaten, including the soft seeds of the tall spikes of the grass trees, and the unripe grain of maize and sorghum, while their liking for fruit sometimes leads to damage in orchards.

Scaly-breasted Lorikeets, like other members of the family Loriidae, have a remarkable adaptation for feeding from flowers. Their tongues are brush-tipped, so that they can open their short bills wide over the open cup flowers of eucalypts to lick out the pollen and nectar.

Like many nectar-seeking birds, and especially lorikeets, Scaly-breasted Lorikeets are nomadic — their wanderings governed by the flowering of trees and shrubs. In the northern part of their range, where they are numerous, their nomadism usually shows up as some fluctuation in numbers, but in the south these lorikeets may be absent for long periods whenever there is insufficient nectar to sustain the flocks.

In common with many other lorikeets, they are gregarious, travelling and feeding in noisy flocks. Their flight is swift and direct, their orange-red underwing coverts giving a flash of colour as they dash through the treetops. But when they land, this colour is hidden and these green-plumed birds vanish among the foliage. Their presence is only revealed by their continuous raucous screeching and chattering.

The nesting season of the Scaly-breasted Lorikeet is variable, triggered or inhibited by seasonal conditions of abundance or scarcity of flowering. The site chosen is often at considerable height, a hollow in a tree trunk or limb, where the two or three eggs are laid on the powdery wood-dust at the bottom of the hollow.

Eastern Spinebill *Acanthorhynchus tenuirostris*

The Eastern Spinebill brings animation and colour to the backyard — busily raiding flowers for their nectar and zealously pursuing flying insects.

In flight the Eastern Spinebill attracts attention with a soft, but distinctive, "flop-flop-flop" of its wings. Conspicuous with the flash of white tail-tips, in its busy and erratic darting between flowers it brings to life the lower levels of gardens or bushes. Its cheery, sharp "chip-chip-chip" and longer, clear, high and musical "cheerwit-cheerwit cheerwit" are delightful sounds to have in any backyard.

As flower and insect populations build up, bush birds generally direct their energy to courtship and nest building. For the spinebills, this is a time of frantic and noisy endeavour — the birds eagerly chasing each other about while calling continuously.

The Eastern Spinebill's nest is a small, neat, deep cup of grass and fine bark strips, well lined with feathers and fine soft plant fibres, all well and thickly bound with webs and suspended by its rim in a slender fork.

As a nectar-seeking bird, the Eastern Spinebill has evolved together with its native food plants over thousands of years. Many native plants, especially grevilleas with their narrow tubular flowers, have sought to restrict access to other creatures that rob nectar but do not carry pollen. The narrow tubes of these flowers require an equally long, fine bill to reach their nectar. Typically, those flowers are red, orange or yellow, standing out strongly against the green of the bush.

Eucalyptus trees have more of an "open cup" design, one which still produces nectar but is also accessible to short-billed birds and insects, some of which are eaten by honeyeaters. Indeed, young, fast-growing spinebills in their nest are fed mostly with insects.

A garden that reproduces the Eastern Spinebill's natural habitat, and where similar food plants are grown, is more than likely to be popular with them and other honeyeaters. Small flowering shrubs grown beneath taller sheltering trees, such as grevilleas, banksias and bottlebrushes, and not only species native to south-eastern Queensland, but also those from much further afield, can be cultivated to attract them.

name: Eastern Spinebill *Acanthorhynchus tenuirostris*

length: 14–16 cm

looks: small; long, fine downcurved bill; red eye; white tail tips; male black through eye; cinnamon throat edged white; female all buff beneath

voice: sharp, long series of "chip-chip" and clear, rollicking "cheerwit"

habitat: heathlands, open forests and woodlands, with understorey of flowering shrubs (especially banksias, bottlebrushes); feeds on nectar and small insects captured at flowers and taken in flight

nesting: small, neat cup of bark strips, grasses, bound with webs, lined with plant down; suspended in foliage of tree or shrub

abundance: common

Shore & Island Birds

The segment of the Queensland coastline that is included in this section extends from the Gold Coast at the New South Wales border in the south, to Cooloola in Great Sandy National Park, just north of the Sunshine Coast.

In contrast to some coastlines of high ranges, cliffs and rugged islands, the Greater Brisbane region may not be the most spectacular of coasts. Fortunately for marine birds it does contain several differing habitats that greatly increase the number and diversity of species that might be seen there.

This coastal landform has been built up rather than worn away by wind and waves (as is usually the case). Shaping much of the length of the Greater Brisbane region's coastline, from Stradbroke and Moreton Islands in the south, to the long curve of golden sand along Cooloola Beach to the north, there has been a steady, shifting process over time. The shape of each long sweep of beach along this coast is caused by wind and waves from the south-east.

The resulting landscape is so picturesque, and has such a rich diversity of flora and birdlife, that many sections have been set aside as national parkland. Within the overall "marine" area, coastal birds are spoilt for choice of habitat. This includes ocean beaches, waters and surf; islands; sheltered bays and estuaries; rocky headlands and reefs, mudflats, sandbars and mangrove swamps. The mangroves are confined to shores of bays and inlets on the sheltered side of islands.

The term "mangrove" is generally applied to both the individual mangrove tree species and a type of vegetation and habitat: trees or bushes that usually grow below the high tide line of coasts and estuaries. All mangroves are well adapted to salty and waterlogged soils and the frequent inundations of rising and falling of tides. They have modifications that allow them to "breathe" in the oxygen-deficient soil, which supports them in the soft mud. Buttressing stilt roots brace the trees, anchoring and supporting them against waves and winds that can reach cyclonic strength.

Mangroves are used by many birds that forage or wander beyond the mangrove forest. Ospreys and White-bellied Sea-Eagles, for example, may nest on mangrove trees but hunt across open waters. On the other hand, there are birds that keep close to the mangrove swamps. These are usually shy and secretive, and difficult to see. The shoreline is rich in birds and bird sighting opportunity.

TOP BIRDWATCHING sites

MORETON ISLAND

This large, almost entirely wild island has several lakes and is home to more than 180 bird species. Egrets, terns and the White-bellied Sea-Eagle can be seen around the shores, while honeyeaters are well represented in the interior. Access is by ferry, and the local tracks are suitable for 4WD vehicles only.

BRIBIE ISLAND NATIONAL PARK

This island is separated from the mainland by the narrow Pumicestone Channel, where tidal mudflats are an important feeding ground for migratory waders. This is a popular site for those seriously interested in shorebirds and migratory waders.

MORETON BAY MARINE PARK

The park is a landscape of islands, beaches, rocky reefs, corals and seagrass beds and extends some 125 km from Caloundra in the north, to the Gold Coast in the south. A destination for thousands of migratory birds, it is estimated that over one quarter of the world's shorebirds migrate along the east coast of Australia.

WYNNUM MANGROVE WALK

Situated on the shore of Moreton Bay at Wynnum, in Elanora Park, the Mangrove Boardwalk also has a hide from which birds may be observed. A typical visit for this site, at high tide, includes some 55 species.

Left: Osprey.

Great Cormorant *Phalacrocorax carbo*

name: Great Cormorant *Phalacrocorax carbo*
length: 80–85 cm
looks: large all-black cormorant with yellow face and throat
voice: usually silent; at nest makes creaky-door sounds, barking and hissing sounds
habitat: large expanses of sheltered waters, most common on coastal estuaries
nesting: in colonies, often with other large waterbirds; builds large stick nest on shrub, reeds or ground
abundance: common and widespread

Australia's largest cormorant is a strongly aquatic species, occasionally seen perched on rocks, posts and jetties — often holding out its wings to dry. Along the coast it may be seen on rock platforms and beaches. Roosting and resting sites include offshore rock stacks and cliffs. The Great Cormorant has often been recorded on Brisbane's Moreton, Stradboke and Bribie Isands, as well as at many sites around Moreton Bay.

Adult plumage is glossy black with a small area of bare, dull yellow skin on the throat and face. When this bird is breeding, its plumage also features white edging on the yellow and white patches on its flanks. Male and female birds have a similar appearance.

The Great Cormorant soars to gain height, holding its wings stiff and straight in a "flying cross" shape. Groups travel straight and level in long lines or "V" formations with steady wingbeats, all birds roughly in unison, gliding together at about the same time.

In Australia, the Great Cormorant prefers mainly freshwater habitats, favouring waters at least several metres deep. It feeds by diving for a wide variety of food species, including many types of fish, freshwater crayfish, crabs, frogs and some large insects.

The species is nomadic in Australia, with many birds moving to inland regions where heavy rains have filled lakes and river pools. There are periodic population explosions when cormorant numbers have built up and birds are forced to relocate once their vital but temporary wetlands have dried out. These birds must then move en masse, usually towards wetter coastal regions and join smaller local populations that would otherwise remain stable over the years.

The Great Cormorant appears to breed any time of the year when conditions are favourable (with water levels high and its food supply assured). It nests in colonies, and commonly in company with other large waterbirds. Colonies can be found in a variety of sites — in trees, on rocks, among reeds, on bare ground or on artificial structures. The Great Cormorant's nest is a large rough platform of sticks, containing three to five eggs. The eggs are white with a faint blue or green tint.

Pied Cormorant *Phalacrocorax varius*

The Pied Cormorant is a handsome, neatly plumed bird with black upper surfaces from head to tail and a white underbody. This cormorant's breeding plumage is more precisely identified when it displays a conspicuous black patch on its flanks — one immediately above each leg. Also on show during breeding, but visible only to a close observer, are very small areas of bare facial skin in pale yellow, blue and pink.

The Little Pied Cormorant, the only other local cormorant, does not have a black thigh patch or the facial colours of the larger Pied Cormorant. If both species are seen together, the Little Pied Cormorant is noticeably smaller, but this might not otherwise be evident. In both species, the male and female are similar. When observed in flight from below, both these cormorants have a similar plumage pattern — a white body with black wings and a black tail. Such plumage also applies to the Black-faced Cormorant, which is found much further south. When travelling in their flocks, Pied Cormorants form a "V" pattern. Their flight is strong and direct, the wingbeats slower than the Little Pied Cormorant's, with heads held slightly higher and the neck often kinked.

The Pied Cormorant is common along the coast and extends inland on major rivers. It can appear in large numbers on some inland lakes and swamps. This species takes mostly fish, but also molluscs and crustaceans. In contrast to the Great Cormorant, the Pied Cormorant sticks more to its home territory — except young birds, which, in their first year, may wander along the coast and venture several hundred kilometres up major rivers.

Breeding varies with local conditions — some colonies are only active from May to July, others also nest in spring. In coastal colonies, nests tend to be close to the ground on the tops of mangroves or other shrubbery. Cormorants of inland swamps make use of taller trees, and their colonies may differ from coastal colonies by including the company of other cormorant species, as well as various egrets, spoonbills and ibis. Nests are large, solid platforms of sticks, seaweed and other debris, in which two to four eggs are laid. These are faintly blue with an overlying chalky layer.

name: Pied Cormorant *Phalacrocorax varius*
length: 70–80 cm
looks: back greenish-black; underbody white with black thigh patch; yellow blue and pink on face
voice: usually silent, but cackles and grunts in nest colony
habitat: rather open and quite deep waters, both inland and marine
nesting: in colonies; large stick nest on ground or broke-down top of a shrub, or low in mangrove swamp trees
abundance: common

Eastern Reef Egret *Ardea (Egretta) sacra*

name: Eastern Reef Egret *Ardea (Egretta) sacra*
length: 60–65 cm
looks: mid-sized egret but shorter yellow legs; overall white or dark grey; eye yellow, bill yellow to grey
voice: harsh, abrupt "yowk" in alarm, otherwise silent, except at nest, gives deep croaks
habitat: shores, mudflats, mangrove swamps, reefs
nesting: mainly Sep–Jan; builds stick nest in secluded site in mangroves, or colonies on islands in north where common; builds on stilt roots, on ground among shrubs and rocks
abundance: common in far north, less common further south

This species is a distinctive egret of shoreline habitats — the rocky wave-washed strip between high and low water, tidal river inlets, mudflats and mangroves, and it is a specialist in the use of this environment. While rocky shores and reefs are its most common habitat, it is also often seen on the intertidal areas of estuarine mudflats, sand and shingle beaches, as well as the muddy or sandy shores and bars of tidal rivers, creeks and mangroves.

The Eastern Reef Egret has two distinct colour "morphs" — white and dark slate grey. Such plumage is genetically determined — the white is more common in the tropics, the grey prevalent across southern Australia.

In size and colour the white form of the Eastern Reef Egret is most likely to be confused with the Little Egret, but it has a much more solid, squat appearance, with legs that are noticeably shorter and thicker. Similarly, the dark form is not as tall and slender as the White-faced Heron, which also has grey plumage, is somewhat paler and has a white face.

When hunting, the Eastern Reef Egret employs its own techniques, suited to its environment. It skulks along the water's edge where the waves surge in over rocks or sand, filling rock-holes and sweeping about masses of seaweed. Here, the Eastern Reef Egret crouches low, and seems to creep or stalk along on its (suitably) short legs. When a small fish, crab or other marine creature is sighted, the bird lunges forward. If the rock pool is rather deep, the bird can be carried partly underwater by the momentum of its strike. At other times, where there is much seaweed in a pool, it may stir and rake through to discover any hidden prey.

The Eastern Reef Egret has few calls, just an abrupt, harsh "yowk" when suddenly alarmed and, during courtship, a deep, abrupt, rather guttural, almost frog-like "yrok, yrok".

For its nest, the Eastern Reef Egret builds a platform of sticks, lined with seaweed. This may be in a fork or supported on the stilt roots of pandanus trees or mangroves, on a rock ledge or on the ground among rocks or low coastal scrub.

Beach Stone-curlew *Esacus neglectus*

The Beach Stone-curlew is a bird of strange appearance and movements; an inhabitant of the fluctuating strip of mud, sand or shingle between low and high tides. It is rather secretive and not often seen. Most sightings are by day, when the Beach Stone-curlew may be seen standing about with little real activity, then occasionally walking forward to jab at some small creature in the mud.

Beach Stone-curlews are usually seen in pairs. They move with slow, deliberate steps and pause motionless for long periods watching any intruder into their territory. They move ahead of any observer approaching along the beach but, if too closely pressed, will take off (showing their distinctive outer wing pattern of bold black and white markings) before landing further down the shoreline. In these circumstances, considering the rarity of the species, it is better to turn back before putting the birds to flight. They should never be excessively disturbed.

The Beach Stone-curlew's habitat includes not only the shores, but also the mudflats of mangrove swamps, where it takes refuge at times. Like the Bush Stone-curlew, this bird is most active from dusk to dawn and most vocal at night, with similar, though harsher calls — a wailing "weer-liew.", repeated about five or six times. During the day, even at a distance, these birds seem to become anxious, giving a nervous-sounding, rather soft "chwip", repeated steadily at intervals of a few seconds.

The nest is a shallow depression scratched out in sand or shingle not far above the high tide mark, and is often among seaweed or other debris. The single egg is heavily blotched and spotted in buff and brown tones when laid, while the chick is also well camouflaged with black strips across its otherwise grey-buff down.

This is a shy species, vulnerable to disturbance on beaches, and consequently unlikely to be seen where there is much human activity. The Beach Stone-curlew occurs only in coastal environments that are mostly undisturbed, inclusive of coastal islands and reefs, islets and spits in estuaries, and beaches where there are mangroves or estuaries nearby.

It occurs sparsely around Moreton Bay. In the Gold Coast area it can sometimes be seen at Southport Spit and on South Stradbroke Island.

name: Beach Stone-curlew *Esacus neglectus*
length: 54–56cm
looks: large shoreline bird; heavy bill; black head with white brow line; bold black and white lines across folded wing
voice: during day, as alarm or when anxious; between birds, a short "chwip" repeated regularly; at night, loud wailings
habitat: in and close by marine tidal zone on beaches, estuaries and mangrove swamps
nesting: shallow scrape in sand, just above high tide line; single egg camouflaged with spots and lines
abundance: uncommon to rare, vulnerable to disturbance of beaches

Common Greenshank *Tringa nebularia*

name: Common Greenshank *Tringa nebularia*
length: 30–35 cm
looks: tall, quite large, alert, upright wader; bill quite long, almost straight; long yellow legs
voice: loud ringing alarm call "tiew-tiew-teiw"
habitat: in almost every inland wetland, permanent or temporary, sheltered coastal bays, estuaries or inlets where there are muddy shallows
nesting: migrant, breeds in Eurasia
abundance: common

This tall, slender and highly nervous wader is usually the first bird to sound an alarm — a loud clear ringing "tiew-tiew-tiew", then takes flight, showing a conspicuous white rump patch as it flies away. Most other species seem accustomed to its alarm call and keep on feeding unless put to flight by a more obvious intrusion.

Seen from a distance, the Common Greenshank forages briskly — stop-start running antics with sudden dashes and changes of direction, spinning around, chasing small creatures flying or jumping up from the wet sand.

They live alone or in small groups or large flocks, often in company with other waders. Greenshanks live in muddy, rather than sandy feeding grounds in a wide variety of coastal and inland habitats: bays and mudflats, mangrove swamps, shallows of harbours, tidal lagoons and occasionally rocky tidal shores. Away from the coast these birds use both permanent and temporarily flooded wetlands, including river billabongs, claypans, flooded irrigated crops and saltworks.

Red-necked Stint *Calidris ruficollis*

name: Red-necked Stint *Calidris ruficollis*
length: 13–16 cm
looks: the most common of several similar tiny waders; in Australia non-breeding plumage has no red on neck; rushes about mudflats with rather hump-backed profile, jabbing short straight bill downwards
voice: very high sharp "chir-it" or "chrit"
habitat: tidal beaches, mudflats, inland freshwater lakes, temporary floodwaters
nesting: summer migrant, breeds in Northern Hemisphere
abundance: common

The Red-necked Stint is the most abundant of the many very small migratory waders. Although named for the rufous around its neck, the colour is generally found on birds at their breeding grounds in the Northern Hemisphere and is so not widely seen in Australia.

This extremely small wader is highly sociable, living in large flocks entirely made up of its own kind, or more often intermixed with other small birds of the shores and mudflats. (Flocks of hundreds or several thousand birds are common.) In flocks the stints keep contact with a quick, high "chrit" or "prip" sound.

In mixed flocks the bird can be recognised by its profile as it moves across sand or mud, rapidly jabbing down with its fine, slightly downcurved bill. It has a distinctly rounded or humped back shape, feeding with the head slightly lower than the back. The many other similarly tiny waders keep their heads slightly higher, with a more distinct hollow seen between the nape and back.

Bar-tailed Godwit *Limosa lapponica*

This is the most prolific wader along the Queensland coast, with over 7000 birds in some places. These large waders migrate annually from far northern Asia, arriving en masse on Australia's northern coast around September each year, then spreading out to the southward coast.

One of the easiest shorebirds to recognise, this bird has a most distinctive bill: very long and curved upwards. The bill's base is light pink, becoming darker towards its black tip. The short tail is cross-barred with bold black lines, usually partly hidden by the tips of the bird's folded wings. In flight, they show a whitish rump and barred tail. Flocks may fly closely together, turning in unison, or else stay strung out in lines, at times low over the water.

Godwits are seen in pairs or small groups on coastal mudflats, sandbars, shores of estuaries and salt marshes. They feed on wet sand and mud exposed by the receding tide, detecting molluscs, marine worms and invertebrates beneath. They also wade out to the depth of their legs to work the deeper water for nourishment.

name: Bar-tailed Godwit *Limosa lapponica*
length: 37–39 cm
looks: quite large wader; tall, with extremely long, almost straight, bill that is pink at base but increasingly black towards tip
voice: in flocks its contact calls is a sharp "krek" or "kak"; in alarm, it is more of a "kirrik" sound
habitat: coastal mudflats, shallows of estuaries, salt marshes
nesting: summer migrant from northern Asia, Siberia, Alaska
abundance: common

Red-capped Plover *Charadrius ruficapillus*

This tiny shorebird sports a distinctive cap that is chestnut rather than red. On the female this is reduced to a faint rufous tint that is mostly on the forehead and sides of the neck.

Red-capped Plovers are usually seen alone or in pairs. In autumn, after breeding, plovers gather into small flocks, often sharing the company of other shorebirds.

Red-capped Plovers are typically seen on the beaches of salt lakes, estuaries, wide open mudflats, sandbars, and occasionally along the shores of shallow freshwater lakes. They run fast, darting along the water's edge, their legs but a blur, stopping abruptly to peck at the sand or mud, staying still, then just as suddenly dashing away again.

Their typical feeding is a run, peck, run, freeze, stop-start pattern of activity. Although to the casual eye the sand often seems devoid of all life, these birds keep busy running, pecking at the sand or mud surface, finding small insects, molluscs and worms.

name: Red-capped Plover *Charadrius ruficapillus* length: 14–16 cm
looks: in breeding plumage easily distinguished by rusty red cap on crown and nape; cap present but not as obvious in non-breeding plumage
voice: harsh, abrupt "yowk" in alarm; at nest deep croaks; otherwise silent
habitat: greatest numbers on inland wetlands, salt lakes, river bed sandbars, claypans; also on sheltered marine habitats, such as shores of estuaries and coastal salt marshes
nesting: nests almost any time of year, but usually Sep–Dec; nest is shallow scraping in sand, on beach above high water line, or among gravel or stones of a near-dry river bed; spotted and blotched eggs are difficult to detect
abundance: common

Black-winged Stilt *Himantopus himantopus*

name: Black-winged Stilt *Himantopus himantopus*
length: 33–37cm
looks: wading bird with extremely long, deep-pink legs; long straight needle-like bill; white body with black back and wings; adults black down back of neck; juveniles instead have blackish patches around and above eyes
voice: yapping sound, quite high, abrupt, repeated often between birds foraging across waters of swamp or lake
habitat: shallow freshwater lakes and other wetlands, open waters rather than with vegetation, floodwaters, tidal estuaries and mudflats
nesting: breeds in scattered rather than dense colonies, or as isolated pairs; a shallow hollow in sand on small islets, or mound of vegetation piled up in very shallow water
abundance: common, nomadic

Some of the most beautiful images of birds, whether seen through binoculars or reproduced on film, are those of stilts standing in shallow, mirror-sleek waters. Captured in these conditions, the stilts' long pink legs meet at the surface where their inverted reflections duplicate the graceful poses of these elegant birds. On such still days the yapping calls of stilts seem characteristic of an entire vista — the wide expanse of water and the whole slow-moving parade of birds in their hundreds, stepping in symmetry with their reflections across glassy shallows.

This display can be missed because the Black-winged Stilt is such a common species. Individual birds, and even large groups gathered around lakes, swamps and estuaries may be scarcely noticed. The Black-winged Stilt's marine sites include fresh, brackish and saline waters of mudflats, sandflats, sheltered tidal inlets, harbours and, occasionally, mangrove shallows. However, the Black-winged stilt is probably even more widespread inland on freshwater lagoons, lakes, swamps and temporary flooded outback salt lakes.

Around Greater Brisbane there are few birds with which to confuse this one, except the Banded Stilt (which is recorded only rarely in the region). As the name suggests, the Banded Stilt has a broad, dark brown band across its white breast (juveniles lack this band). On the other hand, the Black-winged Stilt displays a black band down the back of its neck (juveniles instead have a smudge of black around each eye). On the rare occasions they are seen together, the Banded Stilt has a deeper, plumper body shape, its legs slightly shorter than those of the Black-winged Stilt (but still very long).

For all their terrestrial poise, Black-winged Stilts are perhaps even more elegant in flight. Airborne, their immensely long deep pink legs trail far behind white bodies carried on wings that are entirely black. Upon landing in shallow, mirror-surfaced water, those long legs will reach down to shatter their reflected images.

Black-winged stilts nest in relatively small colonies (sometimes only a handful of birds), or as isolated pairs. A small mound of plant material is built up in shallow water — each nest a tiny, individual island.

Australian Pelican *Pelecanus conspicillatus*

This universally familiar, huge waterbird is famed for its massive bill and is often seen resting or otherwise taking it easy while perched on mooring posts, jetties and beaches. In many areas the pelican keeps in close contact with humans — especially those who share the pelican's interest in fishing.

Pelicans also form a familiar image in flight (with heads tucked into shoulders — forming a tubby flight profile — and great wings held out flat) or when gliding down to land (with webbed feet held out and forward, ready to cushion their impact with the water).

When a flock of pelicans initiate take off, commotion ensues. Although pelicans have an extremely light skeleton relative to their overall body weight, each bird seems to struggle to get airborne, with prolonged flapping and kicking. As their speed across water increases, and each bird begins to lift, their feet continue to pedal at the water rushing underneath, leaving a chain of splashes across the surface, until finally there is sufficient air speed to lift clecr.

In flight, pelicans are transformed from lumbering behemoths, to graceful, even majestic, aeronauts. Each takes its position in the flock's overall formation; where they detect the rising air of a thermal and begin to soar in a tight upward spiral, wings held rigidly at full span and rarely flapping. Within the column of rising air they are soon lifted to a height where, despite their size, they may disappear almost out of sight.

When satisfied with their altitude, (or perhaps when their thermal ceases lifting), the flock sets off in a long, sustained glide to their next destination, probably lifting again, cctching thermals and repeating the process many times on long voyages. Pelicans arrive in great numbers this way to distant inland lakes such as Lake Eyre. There they feed on the mass of aquatic life that proliferates in many of those temporary lakes, and also nest on the islands that have formed. A few months later the temporary lakes will have dried out, and when this happens the pelicans, their numbers now greatly increased, will undertake the return journey to coastal estuaries, bays and lakes.

name: Australian Pelican *Pelecanus conspicillatus*

length: 1.6–1.8 m

looks: huge; white with black on wings; massive long pinkish bill with pouch slung beneath; head tucked back in flight

voice: in display and when squabbling, deep resonant croaks

habitat: may be found on almost any waters, from sheltered coastal bays and estuaries, to lakes, rivers and temporary floodwaters

nesting: in colonies cn small islands or sand spits of lakes or estuaries; nest is merely a slight hollow into the ground with sparse lining of scraps of dry vegetation around

abundance: common, nomadic

Brahminy Kite *Haliastur indus*

name: Brahminy Kite *Haliastur indus*
length: 45–50 cm
looks: raptor with white head and breast; rufous back; rufous wings and tail; juveniles mottled dusky brown overall
voice: high, harsh, drawn-out, wheezy, whistling, descending "pei-ir-ah"
habitat: sheltered tropical coasts, especially coasts with islands, estuaries, mudflats, mangroves where it is more of a scavenger than hunter
nesting: bulky stick nest high in a tree not far from coastal waters; on lower mangrove tree standing in tidal waters
abundance: common

Adult Brahminy Kites are unmistakable. The wings, back and tail are cinnamon-chestnut, while the head, neck and breast are pure white. Juveniles may be difficult to recognise, having streaky and pale-edged brown feathers (similar to juvenile Black or Whistling Kites).

Flight often provides clues to identification. The Brahminy Kite usually glides, patrolling beaches and mudflats, with its broad wings (in a frontal view) held flat or slightly drooped or bowed, but with wingtips upcurved.

This is a raptor of tropical coasts — a scavenger and opportunistic predator watching for carrion and any injured prey washed up by waves or left by a receding tide. Widespread throughout South-East Asia, its usual range in eastern Australia extends not far south of Brisbane. It is probably a vagrant south of Sydney.

The Brahminy Kite's nest is a bulky mass of sticks and seaweed, in a tree overlooking or in the water, often on the seaward side of a belt of mangrove trees.

Osprey *Pandion haliaetus*

name: Osprey *Pandion haliaetus* length: 50–65 cm
looks: perched has brown body, wings and tail; white underparts and head; bold dark line through eye; in flight is white beneath, except for barred brown on flight feathers, undertail and possibly on neck collar.
voice: usually silent, but often noisy at and around the nest; call is a drawn-out "pee-ier"
habitat: coastal waters, islands, reefs, estuaries and inland on major rivers
nesting: large to huge stick nest, often on edge of an island cliff or rock islet well above water; otherwise uses a tree site, telecommunications tower or other convenient structure
abundance: common away from areas of close human settlement

Well known for its deep (often fully submerged) plunges below the surface to seize fish, Ospreys also perform repeated and spectacular high-diving aerial displays early in the breeding season.

Osprey pairs use the same nest each year, adding more sticks, seaweed, driftwood and other shoreline debris, until some nests become substantial towers up to 2 m high and of similar width. The nest is typically built on a site with commanding coastal views, often perched at the edge of a headland. Trees are used where there are no cliffs or rocky islets and these nests may be well inland, overlooking rivers, or on coasts where the land is flat. Communication towers and other artificial structures are also utilised as nest sites.

The Osprey's clutch is two, three or (rarely) four eggs. The chicks remain in the nest for 50–60 days. After making their first flights, they are as big as adult Ospreys but will continue to return to the nest to be fed.

White-bellied Sea-Eagle *Haliaeetus leucogaster*

The sight of one of these large eagles soaring overhead presents a true picture of avian majesty. With its snow white head and body the White-bellied Sea-Eagle strikes a stark contrast against the deeper tones of the sky and the bird's coastal background. Adults are unique among Australia's large birds for their plumage colour.

The White-bellied Sea-Eagle soars and glides over inshore marine waters and the broader reaches of rivers, where this bird's distinctive silhouette and hunting skills easily distinguish it from other raptors. It holds its wings in a steeply upswept "V" profile, while its short, round-tipped tail is another telling feature.

The juvenile White-bellied Sea-Eagle does not display the white and grey plumage of the adult, but rather one superficially like the juvenile plumage of the Wege-tailed Eagle — both are plumed in tones of brown.

A closer inspection of the White-Bellied Sea-Eagle may be possible while the bird is perched (as is very often their habit), on a high dead limb overlooking ocean shore or river. The raptor's best aspect, however, is dynamic and often appreciated best from a distance. It is when one of these eagles is hunting that the full effect of its impressive physiology is displayed.

Once a fish is spied just beneath the water's surface, the eagle dives steeply. With its wings partly folded, acceleration increases as the feet are held forward, forming a direct line between the eagle's eyes and the target swimming below. There is an instant of critical levelling and braking before the long legs spear into the water and the eagle's body strikes the surface — the spray momentarily concealing the assault. Immediately, the bird's broad wings, with deep powerful beats, haul both eagle and fish from out of the ocean and bear them away. The fish is held securely by the piercing clutch of the bird's formidable talons.

The White-bellied Sea-Eagle constructs its nest from a huge pile of sticks accumulated over many years. Its nest may be high in a tree overlooking an ocean or river, or positioned at the top edge of a cliff or headland on an island or otherwise overlooking an undisturbed part of the mainland coastline.

name: White-bellied Sea-Eagle *Haliaeetus leucogaster*

length: 75–85 cm

looks: unusual for a large eagle, with white body; brownish-grey on back, wings, inner tail; in flight has white beneath on wing linings; soars on upcurved wings; juveniles mottled brown

voice: harsh, nasal, goose-honking "ank-ank-ank"

habitat: mostly coastal, over reefs, inland waters, shorelines, mangroves, swamps, far inland on major rivers, floodplains

nesting: massive pile of sticks on islet, cliff top or large near-coastal tree

abundance: quite common around most of coastline except near southern cities

Crested Tern *Sterna bergii*

name: Crested Tern *Sterna bergii*
length: 43–48 cm
looks: large tern with heavy yellow bill; black cap crested towards nape; non-breeding has streaked crown
voice: usual call is a raucous, harsh "graaak" or "kirraak"; very noisy around nesting colonies
habitat: beaches, both surf and sheltered; oceanic waters, reefs, bays inland along rivers; occasionally near-coastal saline lakes
nesting: in big dense colonies of thousands on small offshore islands; each nest is a slight unlined hollow in sand; 1–2 sandy, brown-blotched eggs
abundance: common

During the breeding season the Crested Tern displays a short, streamlined, black crest at the rear of its crown. At other times, in non-breeding plumage, its forehead and crown are spotted with white. It's bill is long, slightly downcurved and yellow. This bird is the familiar large tern of bays and harbours, where it often roosts on jetties and boats, and may be seen in mixed flocks with other terns and gulls.

The large size of the Crested Tern (around 45 cm long), lends it a quite grand stature when perched among smaller birds. This is accentuated by its rather large head and long neck. The species can be confused with the Lesser Crested Tern, which is only slightly smaller, and is a paler grey on its back and wings. Another point of difference is a deep orange bill (compared with the citrine-yellow bill of the Crested Tern).

This is a bird of northern seas, with Brisbane only just within its southern limit. The fishing territory used by these gulls includes both coastal and ocean waters. Inshore, this tern uses estuaries, bays, coastal lagoons and also follows major rivers well inland. Offshore, the Crested Tern takes fish from around oceanic islands and deep, open ocean waters far out from land. Fish are taken by plunging from a considerable height and striking the fish beneath the surface. The Crested Tern's flight can be leisurely, on rather lazy beats, but can quickly engage a powerful and swift flight — the long pointed wingtips angled back, the wingbeats deliberate, strong and deep.

These terns can be quite vocal; a raucous "graak" is commonly heard. Their noisiest behaviour is reserved for their nesting colonies and directed at any intruder venturing near their nests. Such transgression causes much screeching from birds overhead, as well as from those on surrounding nests. The usual breeding season is October to January, but varies with locality. There may, at times, be a second brood reared in the same year.

On small islands, nesting colonies often consist of thousands of birds packed together. Each tern's nest is a slight hollow in sand or shingle, where one or two large eggs are laid. These are pale, sandy grey, blotched and spotted brown and black.

Caspian Tern *Sterna caspia*

This very large, gull-like tern is around half a metre in length, and has a massive red bill and a black-capped, somewhat angular-shaped, crested head. Among other terns it casts an imposing presence even when, as per its habit, it rests with its body horizontal and head pulled tightly down.

When alert and standing tall, with its neck upstretched and head held high, it looks even bigger. Only the Pacific and Kelp Gulls are slightly larger, but have red-tipped yellow bills and much darker back plumage.

Bold colours: the crisp contrast between the white of the neck and face, the solid black of its crested cap and the crimson of its bill, make the Caspian Tern quite spectacular when observed in full breeding plumage. In their non-breeding plumage, white replaces black on the forehead and streaks disrupt the otherwise solid black crown. Non-breeding males and females look alike.

The Caspian Tern is typically a bird of coastal habitats — sheltered estuaries, bays, inlets, lagoons, tidal flats, and harbours. It shows no preference for either sandy or muddy shores. When patrolling coastal waters it generally keeps inside the reef line, beyond which lies the open ocean.

In flight, the Caspian Tern is not unlike the largest gulls, (with powerful deep wingbeats), but is swifter and more graceful with its long, slender, backswept wings. It patrols the surf line and inshore waters, flying with its bill pointing downwards. It may turn, or briefly hover, before plunging into the water to snatch a fish or other marine creature.

The Caspian Tern's most common call, often given while hovering overhead, is a rasping, abrupt "owgk" or "kowk". Among breeding colonies on islands there are various other calls between these birds, heard also when there is any threat or disturbance to the colony.

Nesting colonies may range from small to occasionally large and are densely packed. Such colonies are usually gathered on offshore islands in shallow depressions in the sand. The Caspian Tern's nest is a simple depression in the sand, usually containing one to three eggs.

name: Caspian Tern *Sterna caspia*
length: 48–54 cm
looks: huge gull-like tern; large crimson bill; white body; pale grey wings; when breeding it has black crown with short rear crest; non-breeding has black crown smaller, white-streaked
voice: deep, rasping, abrupt sounds "owgk, owgk"
habitat: usually inshore, coastal, not much out beyond reef-line, using sheltered bays, estuaries, and well inland on large rivers and lakes
nesting: in colonies on offshore islands; often small, occasionally large, and close-packed; nest a slight hollow in ground; 1–3 brown-blotched eggs
abundance: quite common, but dispersed rather than in large flocks

Pied Oystercatcher *Haematopus longirostris*

name: Pied Oystercatcher *Haematopus longirostris*

length: 42–50 cm

looks: large black-and-white shorebird; long, straight, rather heavy bright red bill; pinkish-red legs; eye and ring around the eye also red

voice: high, not sharp or harsh but pleasant ringing "quip-quip-quip"; in flight "quip-a -peet"

habitat: beaches and mudflats of both open surf beaches and sheltered waters and shores; also mudflats, margins of estuaries; less often on rocky parts of shoreline

nesting: a shallow depression in beach sand, well above high tide line, usually near driftwood, low dune plant or clump of grass

abundance: usually common, but becomes uncommon where beaches are subject to much human activity

There are now three species of oystercatcher along Brisbane's shores, two of which have long been very well-known birds. By comparison, the third species has only recently been found on this stretch of Australia's eastern coast; it may have been here much earlier, but not recognised.

The Pied Oystercatcher is the best known of the three. It is a bird of sandy beaches and mudflats but also visits the rocky headlands and islets that are home to the second species, the pure black Sooty Oystercatcher.

The third species — the South Island Pied Oystercatcher — is from New Zealand. Its legs are obviously shorter than those of the Australian species (giving it a rather dumpy appearance). It also has a white line near the edge of its folded wing and a longer, slightly finer bill.

Oystercatchers are all large, rather heavily built waders with chisel-pointed bills well suited to prising oysters and other shellfish from rocks or probing deep into wet beach sand for pipis. All species stand and forage with a hunched posture and the bill pointing downwards. Flight is fast and direct, with rapid, shallow wingbeats. These birds are usually solitary or live in small family groups spaced out along the coast — each defending a section of beach that serves as its territory.

The call of the Pied Oystercatcher is high, but not sharp or harsh — a ringing, resonant "quip-quip-qwip" rising in pitch and intensity as an intruder approaches. Other loud, clear calls are given in flight and as alarm calls.

Breeding usually occurs from August to January. The Pied Oystercatcher's nest is merely a shallow-scraped saucer in the sand, well above the high tide mark and often where there are many dune plants, scraps of driftwood, shells and stones. The two to three eggs are buff and heavily marked with brownish spots, which make both the eggs and nest extremely difficult to see.

Ruddy Turnstone *Arenaria interpres*

This small migratory wader is named for its foraging technique, turning over shoreline debris of seaweed and stones in search of tiny crustaceans, sand-hoppers and other small creatures that might be hiding underneath. Turnstones are distinctively patterned and easily recognised — whether bedecked in their colourful breeding plumage or more subdued wintering plumage.

The breeding plumage of this species is easily recognised — the bold markings of its head and breast form a network of black bands. The throat and upper breast are black, while the head, neck and underbody are white, crossed from side to side by three irregular black bands. One band crosses over at the shoulders, the second crosses from neck to ears and around the nape, and the third crosses from neck to eye and around the forehead. There is no other Australian wader with a similar plumage pattern. The Ruddy Turnstone's wintering plumage has a similar pattern of black markings, but the colours are comparatively dull. In flight, the white, black and bright chestnut pattern across its back and wings is both conspicuous and distinctive.

Only the juvenile of this species is somewhat lacking in plumage clues to aid identification. It does, however, have the same stumpy-legged appearance, foraging behaviour and a few similar face markings, and anyone familiar with the general character of adult Ruddy Turnstones should quickly recognise the juveniles too.

Ruddy Turnstones are gregarious little birds, usually found in small groups that rush about prodding, probing and turning. Their short legs often lead to them being described as tubby, dumpy or stocky. The body is actually quite slender, tapering to pointed rear.

The Ruddy Turnstone's voice is a rapid, irregular, weak chattering and fast twittering in short bursts — a high "trit-tit-tit-tit-tit" or "trititititit", fast enough to be a rippling trill. It also gives a clear, whistled "kiew".

name: Ruddy Turnstone *Arenaria interpres*
length: 22–24 cm
looks: small shorebird; bold complex pattern of black lines over head and down to breast; back chestnut; all colours and pattern dull in non-breeding plumage, even more so on juvenile
voice: weak, high twittering, as a rippling trill
habitat: coastal; reefs, rocky wave platforms, beaches (especially with stones and weed debris), mudflats
nesting: breeds in Northern Hemisphere in northern summer
abundance: moderately common

Wet Eucalypt Forest Birds

After rainforests, the most dense vegetation cover is provided by wet eucalypt forests. These are more precisely referred to as sclerophyll forests for their vegetation has foliage that is, when compared with the soft and delicate foliage of rainforest, hard and leathery. Forest trees and shrubbery are able to withstand much more arid conditions than those of rainforests. The timber of the trees is denser than that of most rainforest trees, hence the description of these as hardwood forests and those of rainforests as softwood forests.

Wet eucalypt forests have few tree species; usually one or two species dominate, but often much smaller trees thrive beneath their canopy. However, with wet eucalypt forests widespread through high rainfall areas, and with dominant tree species differing from place to place, the total number of tree species of Australia's wet eucalypt forests is actually quite large.

The transition from forest to woodland is gradual. It is usually difficult to say just where one ends and the other begins. The relative density of cover is one way of differentiating forests and woodlands. In forests, the trees are more closely spaced, and forced upwards in competition for space and light. Forest trees are not only tall, but the length of their boles (from ground to lower branches) is comparatively greater, making up at least half the total height of the tree. Canopy is also a reflection of forest density. With an overhead canopy coverage of 50–70% of the sky, forests have much greater coverage than that of woodlands, but less than the (almost totally) closed canopy of rainforests. Rainfall also helps distinguish "wet" and "dry" eucalypt forests. With heavy rainfall and rich soils, "wet" eucalypt forests occur having a near-closed crown canopy, trees often exceeding 30 m in high, and a lush, shrub understorey.

The wet eucalypt forest habitat has a rich bird population and some species are found only in this habitat. Other species find sufficient similarity between the dense vegetation of wet eucalypt forests and that of rainforests to be able to occupy both these habitats. In the forests, flowering trees and understorey shrubs are important for supporting large numbers of nectar-eating birds. With only a few trees dominating each tract of forest, flowering of the trees in adjoining forests may occur at different times of the year. Great numbers of lorikeets and other nectar-seeking birds move nomadically through forests and woodlands, briefly bringing much activity and noise to the treetops as they seek nectar and insects among the gum blossoms.

TOP BIRDWATCHING sites

TAMBORINE NATIONAL PARK

Although usually noted for its rainforests, Tamborine also has some of the best, most easily accessible wet eucalypt forests, dominated by the flooded gum. This type of forest can be observed on many of the walking trails and along the Tamborine Village road where it passes through the national park at Joalah.

CONONDALE NATIONAL PARK

Situated in the hinterland of the Sunshine Coast, on the Conondale Range west of Maleny, this national park is still a convenient, day trip distance from Brisbane. Within the park are preserved three habitat types, each protecting a large and diverse bird population.

LAMINGTON NATIONAL PARK

Lamington National Park, so much promoted as a rainforest wonderland, also has considerable areas of wet eucalypt forest. Springbrook National Park and Brisbane Forest Park likewise have areas of tall wet eucalypt forest.

Right: Australian King-Parrot (male).

Satin Bowerbird *Ptilonorhynchus violaceus*

name: Satin Bowerbird *Ptilonorhynchus violaceus*
length: 28–34 cm
looks: male overall black with strong iridescent deep blue sheen, bright red eye; female very different, body greenish above, greenish buff streaked brown beneath, brown wings and tail, blue-grey eye
voice: silent except during breeding season when male is very noisy with loud grinding, churring and wheezy sounds, whistled calls, and some mimicry of other birds
habitat: dense vegetation, typically undergrowth of wet eucalypt forests, rainforests, and dense thickets of introduced plants such as lantana, or other dense shrubbery of gardens
nesting: nest built and young reared by female alone; a bowl of sticks among foliage in crown of tree or other clump of dense vegetation, anywhere from 1–30 m high
abundance: moderately common in suitable habitats

The bowerbirds of Australia are famed for their architectural ability. Each species employs an elaborate design (structurally and decoratively) for its courtship bower. Bowerbirds are well known for their use of tools, and for their careful choice of colours when decorating their bowers. The Satin Bowerbird, like several other species, has developed a technique of painting the inner walls of its bower. It "chews" charcoal or rotted wood then, using a small wad of bark held in the tip of its bill, wipes the sticky brownish mess up and down the vertical sticks of the bower walls. This is one of the rare instances of a bird species that deliberately uses a tool to accomplish a task.

It was once thought that bowers, with their neat construction and carefully maintained decorations, were merely playgrounds in which the males amused themselves, apparently out of sheer exuberance. But the bower is designed, together with the frenetic displaying and calling of the male, to entice any female bowerbird within range. Once confined between those strong walls, she will mate.

The most highly developed of the bowerbirds are those that build avenue-type bowers, with two tall walls enclosing a narrow deep avenue (sometimes almost a tunnel). Within Australia these include not only the Satin Bowerbird, but also the Regent, Fawn-breasted, Spotted, Western and Great Bowerbirds.

Each of these species decorates its bower with a collection of colourful objects. The Satin Bowerbird, in keeping with the male's own plumage of deep iridescent blue, collects predominantly blue objects. For the sake of contrast, it also includes some small yellowish flowers.

Male bowerbirds take no interest in nest building or raising young. The female builds a nest of sticks, lined with dry leaves, among the upright forks of a crown in a tree, or sometimes in a mistletoe clump, at a height up to 30 m. The usual clutch is two or three eggs.

The Satin Bowerbird is widespead through eastern Australia's coastal wet eucalypt forests and within the Greater Brisbane region. It is a common and familiar inhabitant of national parks in the ranges encircling the Greater Brisbane area, and a visitor to quite a few large backyards supporting remnant rainforest vegetation.

Common Koel *Eudynamys scolopacea*

The Common Koel keeps to densely vegetated rainforest edges and other high dense foliage of tall trees of rainforests and of wet eucalypt forests, where it is very difficult to see.

The call of this large cuckoo is likely to be heard by far more people than those who ever see the bird. Not surprisingly, it is the far-carrying, mellow, ringing, musical call of "quo-eel" that has given this bird its name.

Usually these birds are silent, wary and elusive, but in the breeding season, while mostly out of sight above and through the uppermost foliage, the males call and display. Occasionally when a male Koel is calling from a high perch that is not screened by foliage, his plumage may be seen in detail. When they are not hidden in natural rainforests, but rather in more open habitat (such as remnant rainforest in the suburbs), these birds might be seen if attention is drawn to their "quo-eel" calls.

Most striking, when both sexes are seen together, is the great difference in the plumage of male and female. The male's plumage is entirely black, but with an iridescent sheen that puts a purplish-blue and greenish sheen across the surface of the black. The eye is brilliant red and all the more conspicuous set in blue-black plumage. The long tail helps give this bird an overall cuckoo character.

In contrast, the female is black only on its head, with the rest of its upper parts brownish-black spotted white, and the underparts buff with brown barring. Her call also differs from that of the male, being a series of three or four shrieking whistles.

The species is further differentiated by a third type of plumage — that of immature birds. The upper parts of juvenile Common Koels, from crown to tail tip, are rufous brown and patterned with white and dark brown bars and spots. The underparts are buff and lightly barred brown.

The Common Koel is a very large cuckoo, and true to the family tradition, lays its eggs in the nests of other, much smaller, birds. The egg placed by the Common Koel in those nests is larger than the eggs of the host species, but of roughly similar markings. The larger cuckoo chick soon pushes out of the nest its legitimate occupants and takes over their combined food supply.

name: Common Koel *Eudynamys scolopacea*

length: 40–46 cm

looks: very large cuckoo; overall black with glossy iridescent blue and green sheen; brilliant red eye; female quite different, upper parts black, with prominent white spots from shoulders, over wings and tail; underparts buff-white, with barring across of dark brown; eye bright red over white cheek line

voice: male has a quite pleasant musical loud "quoe-eel, quoe-eel"; female gives series of shrieking whistles

habitat: rainforest, monsoon forest, dense wet eucalypt forest (especially the margins), leafy trees of woodlands, farmlands, parks and gardens

nesting: parasitises much smaller birds, including Figbirds and friarbirds (eggs are of roughly similar colour and pattern, but much larger); when hatched young cuckoo ejects eggs or young of host

abundance: quite common

Southern Boobook *Ninox novaeseelandiae*

name: Southern Boobook *Ninox novaeseelandiae*

length: 25–35 cm

looks: small rufous-brown owl with pale-edged, dark brown "goggles" around eyes; pale rim forms "X" between eyes; wings pale-spotted, streaked; fully feathered legs

voice: double hoot, sounds rather like "boo-book"

habitat: open forests and woodland, and most other habitats with trees large enough to have hollows for roosting and nesting; preys upon mice and other small mammals, crickets, spiders

nesting: Aug–Dec, uses tree hollow with quite small to large entrance; in branch or trunk

abundance: common, sedentary

This owl may be unexpectedly and surprisingly small when first seen at close quarters. Large, dark patches around the pale yellow eyes, make this bird easy to identify. The dark patches are encircled by pale edges that meet between the eyes, forming a large, pale, diagonal cross between them and above the bill.

The rather musical, double-noted "boo-book" call of this small brown owl is one of the most common sounds in the bush at night. On a still night during the Boobook's breeding season this call may be heard, from far off, repeated with monotonous regularity. Perhaps there are less Boobooks and fewer nightbirds now, because throughout many parts of Australia it seems like the nights are more silent than they used to be.

The call, when heard close at hand, is much more interesting than the simple "boo-book" sound that it is when distant. The mellow, rather musical double hoot has its first note higher, the second lower, and is rather throaty in the lowest-pitched calls. Often a second Boobook will answer, the two birds alternating their calls, the male higher, the female lower, until one or the other eventually tires of the discussion, and goes quiet.

While this "boo-book" sound may be the most common or typical call of the species, there are many others. Indeed it has been claimed that the Boobook has the widest range of calls of any Australian hawk-owl.

In the vicinity of the nest hollow, these owls give excited squealing and trilling sounds. Young in the nest, and after fledging, make very high-pitched, cricket-like trilling sounds when begging for food — a noise they keep up most of the night.

The Southern Boobook occurs not only in south-eastern Queensland, but right across Australia, almost anywhere there are trees, from rainforests to open woodlands, and heathlands where large trees are sparsely scattered.

Trees are a vital ecological feature in Boobook distribution and numbers as their hollows are needed for roosting through the year and during breeding months. The nest is no more than a hollow, the entrance often quite small, but going deep to where the eggs are laid on a base of twigs and leaves.

Brown Goshawk *Accipiter fasciatus*

The baleful glare of its intense yellow eye seems to reflect the fierce determination of this hunter — one of the most feared of all forest raptors. This is a skulking predatory bird, which hides in wait behind screening foliage, watching the comings and goings of small birds in the forest. When one of these little birds is a little careless and is too far out into open air space, the Brown Goshawk will launch its attack. If the intended victim can avoid the first snatch of the goshawk's talons, it may, with desperate twisting and turning, reach the cover of dense vegetation.

Even then, the goshawk might not give up the chase. If it notes where its prey took refuge, it is likely to turn and plunge headlong into the leafy shelter or, if it is a dense woody shrub, reach in with its long yellow legs — its talons grabbing at the small bird cowering as deeply as it can retreat into its thicket.

The Brown Goshawk takes not only birds, but also mammals up to the size of a rabbit, as well as reptiles. Prey that is usually taken weighs up to half a kilogram, and may occasionally weigh as much as one kilogram.

The Brown Goshawk occurs in a great variety of habitats, from dense forests to semiarid woodlands. Everywhere the small birds, especially the honeyeaters, have warning calls that ring through the bush when a goshawk is sighted, and these warning calls are immediately passed on by all other small birds within earshot.

Several other similarly size raptors share this and other habitats with the Brown Goshawk. The Grey Goshawk and the rare Red Goshawk are slightly larger, and share similar hunting techniques.

Often it is only the call of the Brown Goshawk that reveals its presence. These calls are most likely to be interaction between male and female, or young goshawks calling to be fed. The usual call is a high "keek-kieerk-kiierk", rising in pitch. Another call is a rapid, rising "kik-ki-ki-ki", possibly in defence of its nest site.

The Brown Goshawk's nest is a large shallow bowl of sticks, quite high in a forest tree. The female incubates the two to four eggs, the male bringing food to the female and chicks (when they hatch).

name: Brown Goshawk *Accipiter fasciatus*

length: 40–50 cm

looks: mid-sized "hawk"; grey-brown upper parts; throat to tail buff-white, cross-barred chestnut-brown, underwings and undertail as seen in flight similarly barred; glaring yellow eye has overhanging brow ridge that gives scowling expression; legs long, bare, powerful; juvenile overall mottled brownish appearance

voice: high, piercing "keek-kieerk kiierk" and rapid, excited "kik-ki-ki-ki-ki"

habitat: tall wet forests, dry forests and woodlands, farmlands, parks and gardens

nesting: builds a medium-sized stick nest, perhaps 50 cm diameter, often into a horizontal fork, well lined daily with fresh green leaves; nest 8–20 m above ground, quite well concealed in foliage

abundance: common

White-throated Honeyeater *Melithreptus albogularis*

name: White-throated Honeyeater *Melithreptus albogularis*
length: 13–15 cm
looks: in SE Qld one to three species of similar honeyeaters, all with olive-green back and white crescents around nape with white underbody; White-throated Honeyeater is the only species with white right up chin to touch underside of bill, as well as bluish crescent over eye; others have either black chin or red over eye
voice: sharp "tsip" as contact call, rasping peevish "queerk", and a soft piping-whistled song
habitat: forests, woodlands, rainforest margins, mangroves, watercourse trees
nesting: small, soft cup nest of grass bound with webs, suspended by rim among pendulous foliage at 5–15 m high
abundance: common

This small greenish coloured honeyeater is one of six similar species, and all but one of them (the Brown-headed Honeyeater), have a distinctive white crescent around the nape of the neck. Four occur in south-eastern Queensland and, as they differ only in very small details, binoculars will probably be needed.

The White-throated Honeyeater has a pure white throat, right up to the base of its bill. The very similar Black-chinned Honeyeater has a black patch from the base of the bill a little way down onto its throat, while the White-naped Honeyeater has a small crescent of bare skin over its eye (the White-throated has pale blue or white).

Foraging for nectar and very small invertebrates is done at woodland or forest heights in foliage and tree crowns.

Breeding can occur at almost any time of the year when there is strong flowering of forests, but usually occurs from July to December. The small, deep cup-shaped nest of fine grass and bark fibres, bound with webs, is suspended by its rim from pendulous parts of upper foliage at 5–15 m.

Yellow-throated Scrubwren *Sericornis citreogularis*

name: Yellow-throated Scrubwren *Sericornis citreogularis*
length: 12–14 cm
looks: yellow throat; black "mask"; long white and yellow brow line
voice: cheery whistle, high and sharp alternating with low notes; accomplished mimic
habitat: eucalypt forests, woodlands, mallee
nesting: domed nest of grass in a narrow tunnel dug into earth bank
abundance: common

Yellow-throated Scrubwrens are usually seen in pairs foraging on the forest floor. Surprisingly, their bold markings are inconspicuous in their natural surroundings, merging seamlessly into the rainforest's background.

Like other scrubwrens, the Yellow-throated Scrubwren harshly scolds any intruder, particularly if the little bird has a nest in the vicinity. In contrast, their song is a pleasant and lively warbling, interspersed with imitations of the calls of a great variety of other birds.

For so small a bird, the nest is enormous. A basic nest is a mass of interwoven black rootlets, palm fibre, dried leaves and black threads of "horse-hair" fungus, with a roof and hooded entrance hole. But because these scrubwrens often raise several broods in a season, each time adding a new "room", many nests often have several egg-chambers, each with its own entry. Consequently, these nests increase in bulk and, being conspicuously placed, can hardly be missed.

Satin Flycatcher *Myiagra cyanoleuca*

The satin-blue sheen across the male's back is iridescent, its plumage deep glossy blue-black, except for lower breast and undertail, which is white. Female plumage is equally beautiful, yet different; from chin to throat to upper breast is deep orange buff. Active in mid to upper levels of dense forests, fluttering about foliage, flycatchers dart out to snap up flying insects. Perched, they are energetically restless, swinging tails from side to side.

Each autumn Satin Flycatchers move north, mostly along a coastal strip to north-eastern Queensland and through to Papua New Guinea and other northern islands. Returning in spring, they find breeding sites in the heavy, high rainfall forests of the coastal ranges of south-eastern Australia, and south of Brisbane. Near-identical in appearance, separating females of Satin and Leaden Flycatchers is difficult. However, the habitat in which nests are found is a useful clue. Preferred by the Satin as nest sites are moist gullies with dense undergrowth beneath the tall trees of the tall wet schlerophyll forests; the Leaden uses much drier open forest as its breeding habitat.

name: Satin Flycatcher *Myiagra cyanoleuca* length: 15–17 cm
looks: one of several very similar species; only other where male has black upper breast and white belly is Leaden Flycatcher, Female Satin Flycatcher has orange-buff breast
voice: repeated sharp, metallic "chwee-ip"; faster higher "chee-ip" and harsh, grating "grzzz-urk"
habitat: in breeding season, moves into wet dense undergrowth of wet forests; with fledged young into woodlands, heaths, mangroves; avoids rainforests
nesting: deep cup of fine strips of bark, bound with cobwebs, camouflaged externally with bark; placed in horizontal fork of dead twig under foliage, 5–20 m above ground; often several pairs not far apart
abundance: summer breeding migrant into SE Australia

Grey Fantail *Rhipidura fuliginosa*

In the wild and encouraged by a few squeaky sounds, this bird will come close and flutter around a visitor entering its patch of bush. In many suburban gardens it is the bird most likely to be dancing and chattering cheerfully about, close to wherever there are gardening activities under way.

Probably one of the most entrancing of the Grey Fantail's abilities is its nest-building skills. Built in a neat cup shape, about the same size as an egg-cup, the nest has a long tail that hangs some 7–10 cm below the nest. Built of fine grasses and bark strips, it is bound together with so much cobweb, that its walls have a felt-like softness and flexibility, while the pale grey, smooth outer surface looks more like dead wood than nest.

There has understandably been much conjecture as to the purpose of the nest's tail; possibly it serves to drain rainwater from the nest. Probably the fantails themselves do not know, but simply follow some inbuilt instructions by which every Grey Fantail, coast to coast, rainforest to semi-desert, follows that same plan.

name: Grey Fantail *Rhipidura fuliginosa* length: 14–17 cm
looks: small active bird (like small Willie Wagtail); tail very long, almost constantly widely fanned and waved side to side; upper parts mid to dark grey; tail white-tipped; white streaks on brow, behind eye; underparts creamy-white, with dark band across breast
voice: cheery outpourings of squeaky, scratchy sounds
habitat: most vegetation types, from mangroves and rainforests to dry woodlands, parks and gardens
nesting: tiny, neat cup with long slender tail
abundance: common

Australian King-Parrot *Alisterus scapularis*

name: Australian King-Parrot *Alisterus scapularis*
length: 42–44 cm
looks: large parrot; bright crimson body; vivid green wings; female has pale green head, yellowish face to neck, crimson breast to undertail coverts
voice: in flight a sharp, strong, clear "krassiek" and "charrak"; in alarm a harsh, metallic screech
habitat: tall wet eucalypt forests, woodlands, rainforest margins, mangroves
nesting: a deep hollow high in a tall forest tree
abundance: common where sufficient forests remain

These large and spectacular parrots inhabit the forests of both coastal lowlands and mountain ranges. There they keep mostly to rainforests and the heavily timbered country of wet eucalypt forests, but also visit nearby open woodlands, orchards, parks and gardens. In some areas this is quite common, and large numbers may gather where fruit or seed become plentiful on trees or undergrowth.

Most of their feeding is done in trees and shrubs, where they take the seeds of eucalypts and wattles, as well as fruits, berries, nuts, and blossoms (for their nectar and pollen). Occasionally Australian King-Parrots feed on the ground, where they find fallen seeds of grasses and shrubs, and occasionally visit farms to pick up spilt grain.

In the bush, the behaviour of Australian King-Parrots sets them somewhat apart from other parrots. Their flight, even seen at a distance, is much more direct and steady — most unlike the buoyant, undulating and sometimes quite erratic flight of the smaller parrots. Being considerably larger, they are slower and more deliberate in their movements, more wary, and harder to approach.

As befitting a bird of such magnificent plumage, the courtship display is well developed, with both sexes making full use of their bright colours in their actions and responses. The male Australian King-Parrot puffs out the scarlet feathers of his head, flicks out his dark green wings to display the intense colours of his shoulders, and gives shrill calls. In response, the female fluffs out her head plumage, and solicits courtship feeding by calling and bobbing her head.

The Australian King-Parrot chooses a deep hollow high in a forest tree for its nest, often in the trunk rather than branches. Such hollows can be very deep, and although the entrance may be 10–15 m above ground, the floor of the nest chamber may be almost back at ground level inside the base of the trunk.

When the three to six eggs are laid, all incubation is performed only by the female. The male remains in close attendance, however, feeding her at intervals by day, and roosting nearby at night.

Crimson Rosella *Platycercus elegans*

Of Australia's many parrots known as "rosellas", the Crimson Rosella is probably the most richly plumed. A flock of these brightly coloured birds feeding on the forest floor, or passing overhead in swift undulating flight between the tall straight trunks of the forest trees, is one of the most spectacular sights of tall wet forests.

Rosellas all have well defined cheek patches, and the Crimson Rosella has cheeks of bright, violet-blue. Adults are so boldly plumed that identification should not be a problem. Juveniles are a nondescript green. These young birds form their own small flocks, rather than associating with adults, and the lack of any mature birds in these flocks does not help with identification. All rosellas are long-tailed, medium-sized parrots that feed largely on the ground. They feed on grass seeds, as well as tree and shrub seeds on the ground. In the tree foliage, these birds forage for berries, nuts and fruit.

Unlike the nomadic lorikeets, these rosellas are mostly sedentary. They use a wide variety of habitats, from coastal lowlands to the highest ranges, and are abundant in Lamington National Park on south-eastern Queensland's high McPherson Range.

The Crimson Rosella is one of the most common parrots of urban areas of Brisbane city. It is found in tropical and temperate rainforests, in the tall wet eucalypt forests and fern gullies of the coastal slopes of these ranges, as well as in open eucalypt forests, woodlands, farmlands and coastal scrubs. Usually Crimson Rosellas will be seen in pairs or small groups of up to five or six birds, although immature birds may also be seen in slightly larger flocks.

Crimson Rosellas nest in spring and early summer. The male has a courtship display where he droops his wings, jerks his widely fanned tail side to side, and chatters loudly and continuously.

The nest is made in a hollow of a living or dead tree, or in a tree stump, and may be either quite low or very high. The four to eight white eggs are laid on wood dust at the bottom of the hollow and incubated by the female for about 20 days. The young leave the nest at about five weeks of age and remain dependent on the adults for a gradually decreasing portion of their food for a few weeks longer.

name: Crimson Rosella *Platycercus elegans*

length: 32–37 cm

looks: medium-sized parrot; body entirely crimson except for deep blue cheek patch; wings deep blue above and beneath except brownish main flight feathers; tail deep blue except white tips

voice: clear, ringing "k-tee-i-tip", the "tee" part loudest

habitat: tall wet dense eucalypt forests, rainforests, partly cleared land near heavy forests, parks and gardens with, or near, or similar to these vegetation types

nesting: hollow of a tree at variable heights

abundance: abundant, including some parks and gardens of cities

Spotted Pardalote *Pardalotus punctatus*

name: Spotted Pardalote *Pardalotus punctatus*
length: 8–10 cm
looks: short-billed; stumpy-tailed; upper parts heavily spotted, except red rump; underparts buff except male has yellow throat
voice: loud, sharp, vibrant screeches; softer chatterings when feeding
habitat: eucalypt forests, woodlands, mallee
nesting: domed nest of grass in a narrow tunnel dug into earth bank
abundance: common

A previous common name for this pardalote was "Diamondbird" — an appropriate description of the plumage adorned with so many small white spots. Many of these spots are on dark feathers, so that their sharp points are indeed about as diamond-like as possible on any plumage.

But it is only rarely that the Spotted Pardalote can be observed so closely that such fine detail is visible. These tiny birds feed among treetop foliage, with their stubby bills taking lerps, manna and insects from leaves.

Their presence might only be revealed by their calls — mostly a soft, musical "weep-weeip". Their song is a sequence of clear whistled notes, "whee, whee-bee", often described in words as "sleep, may-bee".

Occasionally foraging pardalotes work their way through low foliage much closer to the ground, or they may be seen near the entrance of their nest tunnel, when a much closer view might be obtained (and the beauty of plumage detail can be fully appreciated).

Outside the spring–summer breeding season, Spotted Pardalotes gather in flocks. Usually these flocks are comprised of 15–30 birds, but are sometimes considerably larger. The flocks wander to forage further and reach other food sources, but most of their movements are over short distances.

In early spring the flocks return and the birds disperse as pairs to their nest territories. The tunnels of Brisbane's Spotted Pardalotes will usually be drilled near the top of an earthen embankment, occasionally into the side of an earth-filled hanging basket. In a larger chamber at the end of the tunnel the pardalotes construct a domed nest, with its side entrance facing the earthen entry tunnel.

It seems somewhat paradoxical that these tiny, jewel-like birds of the high forest canopy, of open air and sunlight, should come down to the ground to scratch out a narrow dark tunnel to raise their young — rather than build a nest similar to those other small birds build amid the foliage of trees. Perhaps this subterranean strategy gives them more choice of territory with abundant food and removes any necessity to compete with other small birds that need tree hollows for nesting.

Eastern Yellow Robin *Eopsaltria australis*

Of all the small birds of Brisbane's backyards, this robin is one of the most familiar. Inquisitive, friendly and may often seem fearless, keeping close by when gardening is in progress, especially where digging exposes the soil and the occasional insect or other small creature as prey.

This robin often perches, almost motionless, on a low branch or the side of a tree trunk, intently watching camping, picnic or backyard gardening activities with keen eye. At the same time it will be alert to any movement, quick to dart down to take tiny prey on the lawn or in leaf litter. It seems to keep a watchful eye on those who enter its territory, but in a way that makes this bird seem a custodian of that place (whether backyard, nature park or bush camp site).

The Eastern Yellow Robin is a bird of diverse habitats. It may be found in rainforests, eucalypt forests and woodlands, dry mallee, swamps and pine plantations. It may live in or visit the gardens of homes in or adjoining any of those habitats. Native and introduced shrubs, even the freshly turned earth of ornamental or vegetable gardens, may all contribute towards providing a backyard acceptable not only to this yellow robin, but to other, more secretive species in suburban yards.

Above all else, the Eastern Yellow Robin has a need for protective understorey shrubs, such as the lower strata of taller vegetation. Such shrubs provide perches from which to scan the ground (whether it is open and grassy, leaf littered or cultivated). A backyard with these essentials, especially if adjoining or near bushland, parks, forests, or other similarly bird-friendly backyards, may well be attractive to the Eastern Yellow Robin. As additional bribery, when one of these birds is nearby, turning or raking soil or compost rich in worms, together with a bird bath in which to splash, is likely to make most suburban backyards more welcoming as a home to this robin.

In such gardens, the Eastern Yellow Robin may even nest in thickets or denser areas of shrubbery. Its neat cup nest, bound with cobwebs and skilfully camouflaged with bark strips, will be tucked into a fork anywhere from 1–5 m above the ground. Often, the presence of a nest is betrayed by the agitated scolding of the robins should any intruder come near. Even then, it is so well concealed that it can be very difficult to see.

name: Eastern Yellow Robin *Eopsaltria australis*

length: 15–16 cm

looks: typical plump robin shape; clings to side of tree to watch the ground; upper surfaces all grey; underparts from throat to undertail coverts bright yellow; from behind, and especially in flight away, northern race shows a bright yellow rump; southern races are more greenish-yellow on rump

voice: clear, even, piping whistle, mellow rather than sharp, but carries a fair distance through the forest

habitat: wet eucalypt forests, dry open eucalypt forests, woodlands, banksia heathlands

nesting: neat, deep cup of grass and bark, heavily bound and covered with webs and with long strips of bark attached around the outside until it looks the same in colour and texture as the branches upon which it is built

abundance: common, seasonally nomadic

Bell Miner *Manorina melanophrys*

name: Bell Miner *Manorina melanophrys*

length: 18–20 cm

looks: smallest of the honeyeaters known as miners; overall olive-green except for rufous tone on wings and tail; bright yellow bill and between bill and eyes; small bright red triangle behind eye

voice: clear, abrupt "peep", which from many birds at once sounds like the tinkling of small bells

habitat: only one of four species using tall wet eucalypt forests, rainforests and dense gully thickets, swamp gum forests, some gardens with similarly dense vegetation

nesting: some breeding activity throughout the year in colonies; small cup nests suspended by rim from twigs of foliage; other birds in territorial groups probably assist with feeding the young at each nest, and certainly in defending site

abundance: common in areas where habitat is suitable for colonies, otherwise absent

There seems to be an unceasing high "tinking" sound filling the cavernous space between high canopy and the forest floor coming from the treetops in forests when Bell Miners are calling.

This sound is from the combined calls (each a sharp, metallic "tink") of a colony of hundreds of Bell Miners high above in the forest canopy. The call is repeated almost continuously by each bird, and by so many birds in all directions, that the listener is completely surrounded, enveloped in this seemingly ceaseless "tinking" noise.

By scanning the high foliage it is occasionally possible to see the movements of small olive-green birds moving far above. Bell Miners inhabit tall wet eucalypt forests, mostly keeping to the high canopy foliage. A colony may have several hundred birds, but in their olive-green plumage at such height it is difficult to get a good look at any individual bird. Bell Miners inhabit tall wet eucalypt forests, typically where the undergrowth is very tall and dense and often along creek lines.

When seen close up, this bird's most distinctive feature is a triangular bright orange patch of bare skin close behind its large dark eye. Together with the yellow bill and orange legs, this gives some colour to a bird with plumage that is otherwise rather drab.

The species is highly aggressive and colonies employ several tactics to deter intruders. Birds support each other, approaching intruders in a long, direct, threatening glide with wings held up in a "V" shape. Within a few seconds there is a swarm of Bell Miners encircling the greatly outnumbered creature that has come into their territory, all giving noisy abuse and threatening postures.

It is thought that by combining forces in this way, living in a colony of hundreds of birds, Bell Miners are able to dominate a site with rich food resources. For the Bell Miners this consists of lerps — sap-sucking scale insects which attach to foliage and have a sugary, protective coating. Bell Miners often maintain their colonies in sites of rich food supply over many years.

The nest is a small, rather flimsy open cup of grass suspended by the rim. It is sometimes located in undergrowth shrubbery but more often in the mid to high foliage of trees.

Golden Whistler *Pachycephala pectoralis*

The male Golden Whistler is one of the most spectacular birds likely to be seen in and about the foliage of backyard trees and shrubs. After its loud ringing calls have attracted attention, the Golden Whistler is usually quite easily found, followed and observed.

Despite the male's spectacular and conspicuous colours, he takes his turn at egg-warming duty, and is often seen on the nest during daylight hours, as if defying the logic that the nest and contents would be safer with the less noticeable dull-plumed female sitting on the eggs by day. But as long as he remains deeply snuggled down onto the eggs or small chicks, the brightest of his plumage (white and gold), remains hidden below the rim of the nest.

Through the breeding season, usually August to January, Golden Whistlers form pairs, but they are otherwise solitary, or else go about their business in small feeding-flocks of mixed species (sharing company with thornbills, treecreepers, robins and shrike-thrushes). While the plain-plumed female tends to keep company with the less colourful company, the male is more inclined to associate with other yellow-breasted birds, such as the Eastern Yellow Robin and Crested Shrike-tit.

To encourage this beautiful bird to visit or take up residence in a backyard, trees and shrubs with abundant foliage are needed — species chosen for their nectar-producing flowers, which attract honeyeaters and lorikeets. While the presence of nectar and insects hovering about flowers or on foliage at certain times of the year is a bonus for Golden Whistlers, they would work over the foliage and bark at other times anyway.

Golden Whistlers construct a deep cup-shaped nest of fine twigs, grass and soft bark, placed in an upright fork. Often this fork has more than two upright stems around the nest, which is supported rather than suspended by the rim amid hanging leaves (like the nests of many honeyeaters). It is well hidden where the twigs and foliage are dense. Shrubs with still, prickly leaves, such as certain dryandras, are favourites.

For foraging, protection against predators, and for a whistler-friendly backyard, trees and shrubs with plenty of foliage are needed. Some dense, stiff and rather prickly foliage is often favoured by these birds as nest sites.

name: Golden Whistler *Pachycephala pectoralis*

length: 16–17.5 cm

looks: male underbody and nape bright yellow; head and breast-band black; throat white; female grey, paler beneath, lemon tint to undertail coverts

voice: loud ringing notes, with increasing intensity, "whit-whit"

habitat: diverse; rainforest, eucalypt forest, mallee, brigalow

nesting: open cup of fine twigs, grass, bark, in dense part of foliage of tree or shrub, 0.5–6 m above ground

abundance: common

Rainforest Birds

Many fascinating and often very beautiful birds are dependent upon the rainforest habitat. For some, rainforest is their only habitat. These are birds so specialised, so adapted to the sheltered, gloomy, damp environment beneath the enclosing foliage canopy of rainforests, that they could not survive except under that sheltering roof of vegetation.

These may be birds of the lowest levels of the rainforest, where sunlight rarely penetrates, foraging by scratching through decaying leaf and twig litter of permanently damp soil, or they may be birds of the high yet shadowy rainforest canopy, seeking native berries and fruits. Only the rainforest has those unique foraging niches; there are few if any similar places in habitats outside the protection of the rainforest.

Most birds that occupy the special habitats within rainforest have no possibility of surviving in any other habitat, and never, even briefly, venture outside. These "specialist" rainforest birds, especially those dependent upon the unique forest floor environment would be unable to find equivalent habitat anywhere other than their specific type of rainforest, or indeed outside a certain niche in that rainforest. These birds usually have very little tolerance to disturbance of their delicate habitat. Wherever rainforest is destroyed, so too are most of its truly unique birds.

Some birds that are largely reliant upon rainforest habitat are the Noisy Pitta, Green Catbird, Rufous Scrub-bird, Paradise Riflebird and Olive Whistler. The rainforests encircling the greater Brisbane area of south eastern Queensland have these five species unique to this region: Rufous Scrub-bird, Albert's Lyrebird, Logrunner, Regent Bowerbird and Paradise Riflebird.

There are some birds, however, that visit rainforests, if only its outer foliage, and feed in many other habitats. Rainbow Lorikeets, for example, may be seen in rainforests, eucalypt forests, heathlands, suburban trees and backyards — wandering wherever trees are in flower.

The Rufous Fantail uses not only rainforests, but also places of dense undergrowth — eucalypt forests, mangroves, and others that have the shadowy or gloomy space beneath a largely closed canopy. But it feeds upon insects on and about the foliage, and similar opportunities can be found in many other habitats. This colourful fantail is a migratory species, and is able to use the many habitats and food sources along its migration route, whether they be undisturbed rainforests or suburban gardens.

TOP BIRDWATCHING sites

LAMINGTON NATIONAL PARK

An encounter with the birds of Lamington's rainforests could best begin soon after sunrise. Along the heights of the McPherson Range, a 21 km walking track may reveal a Rufous Scrub-bird, Noisy Pitta, Rufous Fantail, Pale-yellow Robin, Eastern Whipbird, Green Catbird or Yellow-throated Scrubwren. The Eastern Bristlebird and Powerful Owl also occupy this habitat.

TAMBORINE MOUNTAINS NATIONAL PARKS

For those wanting to experience rainforest and some of its birds much closer to Brisbane, and without such long walks, a cluster of national parks on the small but high plateau of Tamborine are an easier introduction to rainforest trails and their birds. The elusive Albert's Lyrebird makes its home here.

SPRINGBROOK

This national park is set on a plateau south-west of the Gold Coast, which reaches an altitude of 1000 m, and forms an eastern extension to the McPherson Range. The Satin Bowerbird is one species to watch out for.

RAVENSBOURNE NATIONAL PARK

This national park is situated on a spur of the Great Dividing Range north-east of Toowoomba and about mid-way between Toowoomba and Esk. Birds here are attracted to the fruit of rainforest trees. Canopy species include the Superb Fruit-Dove, Rose-crowned Fruit-Dove and Wompoo Fruit-Dove.

Above: Emerald Dove.

Regent Bowerbird *Sericulus chrysocephalus*

name: Regent Bowerbird *Sericulus chrysocephalus*

length: 25–29 cm

looks: male unique; black body; golden crown and shoulders; golden panel across flight feathers for almost full length of each wing; eyes intense yellow; female completely different; overall brownish with pale pointed streaks from nape to upper back, and also underbody; black patch at rear of crown on head; eyes brown

voice: usually silent, except in breeding season, when male is noisy with loud, harsh, wheezy, churring sounds in displays around his bower

habitat: rainforests, wet eucalypt forests, but usually in autumn and winter it moves out of these dark habitats to open forest or woodlands and also into nearby parks and gardens

nesting: female only builds the nest, incubates, raises the young; nest is a rather flimsy cup of sticks, well hidden in dense foliage of dense vine or mistletoe

abundance: generally scarce, locally common in optimum habitat sites

A Regent Bowerbird passes through a shaft of sunlight slanting down from a break in the forest canopy far above; behind him is the darkness of the rainforest. For those few seconds the flying bird appears more like a huge, gold and black tropical butterfly, the black of his plumage lost against the shadowy background, while the brilliant yellow of wings and head is highlighted by the sun's spotlight.

This brief but spectacular image, out of the patch of sunlight, has gone almost before it is recognised, back into the gloom of the rainforest. The Regent Bowerbird must rank high on the must-see list for Australian birdwatchers, not only for its plumage, but also for its bower construction displays at the bower.

Like other avenue-type bower builders, the Regent Bowerbird paints the sticks with yellowish colour from crushed leaves, and collects various objects such as snail shells, berries, leaves and other items of yellowish-brown colour for display.

The male begins his display when a female approaches the bower, prancing about, showing golden colour by drooping his wings and making chattering, wheezing, grating and scolding sounds.

The breeding season is from about October to January. The male mates with the various females that visit his bower, but takes no part in building nests, incubation or feeding of young. The nest is a rather flimsy open saucer of twigs, concealed in a tangle of vines or dense vegetation at 3–20 m height.

Except for their attendance at the bower, Regent Bowerbirds keep mostly in the mid to upper levels of the rainforest, where they feed on native fruits. While adult breeding birds probably keep to rainforests or close to the rainforest margins, at other times of the year flocks of ten to thirty birds gather to wander into nearby open country.

The best chance of seeing one of these truly magnificent birds will be in one of the national parks of the ranges at the southern edge of the Greater Brisbane region, where they come around visitor centres and camping grounds (at Lamington National Park) at certain times of the year.

Pacific Baza *Aviceda subcristata*

The Pacific Baza is a most distinctive bird — both for its hunting style and its appearance. It circles low over and around the upper foliage of trees, partly gliding or with the odd leisurely flap of its large, broad wings. Occasionally it may be seen snatching at foliage, for this is where its prey is hidden. Among the creatures taken are tree-frogs, large insects, lizards and tree-snakes, small birds and nestlings, and some native fruits. This bird also hunts from a high perch beneath the forest canopy, closely watching the surrounding tree foliage.

In appearance also, this species is unique. Apart from its crest, it has wings that are exceptionally broad towards the outer wing, heavily banded across the flight feathers, and with underwing coverts of deep orange-buff. The head and neck are pale grey, the breast and belly are conspicuously banded in brown and white and the tail is heavily barred like the wingtips. A bright yellow eye completes the picture of this colourful and distinctive bird of prey.

Distinctive also are the Pacific Baza's spectacular aerial display flights. The bird climbs with deep, strong wingbeats, plunges down with wings held high (showing underwing colour and pattern), then swoops up to repeat the alternating rising and plunging, all the while calling loudly. Its calls can sometimes attract attention to this species and is a distinctive rising and falling, musical "whiech-yoo, whiech-you" repeated many times.

Breeding usually occurs from spring to summer. The nest can be found in a forest or woodlands tree, often near a stream or billabong. The large shallow bowl of sticks and leaves, lined with leaves, is typically positioned at the mid to upper level of a tree.

The Pacific Baza favours the broken margins of rainforest. Often in regrowth areas, and in parks and gardens, the trees are separated by open spaces, which allow this bird to glide around and among the trees close to their foliage. This bird of prey has been recorded widely across Brisbane's suburbs, and might be a visitor to almost any garden or park with large trees, but especially those adjoining rainforests, near the river or near wetland areas.

name: Pacific Baza *Aviceda subcristata*

length: 34–45 cm

looks: unique small hawk, pointed crest at rear of crown; head and neck pale grey; body rufous with white stripes across breast and abdomen; wings and tail mid to dark grey above; heavily barred black on pale grey for most of wings underside

voice: usual call is a scratchy, but not unpleasant, rising and falling "whieech-yoo", repeated several times

habitat: margins and broken edges of rainforests, gallery forests, swamp forests, woodlands, parks and gardens adjoining rainforests

nesting: large nest of sticks, well hidden in dense part of foliage, and lined with leaves

abundance: uncommon

Green Catbird *Ailuroedus crassirostris*

name: Green Catbird *Ailuroedus crassirostris*

length: 24–32 cm

looks: back, wings, tail all quite intense mid green with clear white spots at tips of the larger feathers; head brownish, merging to dull yellow of underside; head and neck, with entire underparts, are streaked white

voice: rather cat-meowing, undulating, grating, wailing "heer-l-aar"; with familiarity its voice is not unpleasant, becoming one of the typical sounds of rainforests

habitat: rainforests, dense paperbark thickets, occasionally adjoining wet eucalypt forests

nesting: a large, thick-walled, bulky nest of quite large sticks, lined with dry leaves; unusual is a layer of decayed softwood between leaves and base of nest; male assists in the feeding of young

abundance: common

The Green Catbird advertises its presence with unmistakable, catlike wailings — loud and frequent. But even then, like so many rainforest birds, it can be difficult to spy in that dense and often dimly lit habitat. Both sexes give these calls, and also a sharp clicking that is probably a "keeping in contact" sound.

Pairs hold territories for many years, and appear to remain in that locality. Small groups of birds, apparently of young from the last breeding season, wander before setting into a territory of their own.

The Green Catbird's voice is an undulating, grating wail which, as one becomes familiar with it, is so frequent and widespread that it becomes a sound one expects as a kind of rainforest greeting. This call is indeed quite catlike, varying from a short, but very recognisable "heer-i-aar", to a very slow, drawn-out, wavering "heeeir-ieee-aaa-aaar". This call is mostly heard in the breeding season, from September to January, and then much more frequently at dawn and dusk.

These birds are mainly fruit-eaters, foraging at all levels of the rainforest from forest floor to high canopy. There they gather native figs and other fruits, insects, frogs, nestlings of small birds, flower heads and some leaves.

The Green Catbird is confined to rainforests and adjoining dense wet thickets of tall eucalypt forests along the ranges of south-eastern Queensland and north-western New South Wales. A second species, the Spotted Catbird, has similar calls but is found in north-eastern Queensland.

Catbirds are closely related to bowerbirds; however, they do not construct bowers for courtship and form monogamous pairs. While only the female builds the nest and incubates the eggs, the male joins in the feeding of the young and the defence of the nest.

The nest is very large for the size of the bird, with much of its bulk contained in the nest's thick walls. The exterior is made of sticks, bound with vine tendrils, while the inner lining is made of dry leaves, beneath which is a layer of soft, decayed wood pulp. Typically the nest is built into the dense crown of an understorey tree or shrub in a position where there are multiple stems and forks to support its bulk, perhaps in a stinging tree or in the crown of a tree fern.

Rufous Fantail *Rhipidura rufifrons*

In gloomy rainforests and the shadowy undergrowth of tall eucalypts, a bouncing, dancing spot of bright rufous catches attention. A Rufous Fantail is spied through a small break in the forest canopy, flitting from twig to twig and chasing insects about the foliage in fluttering, twisting and turning flight. Its large tail, repeatedly and widely fanned, now and then catches a shaft of sunlight and glows briefly, brightly, as if lit by an inner fire.

When following the fantail's movements it is possible to see it pause briefly, several times, at the same point each time — a woody lump on a slender branch. A closer look will reveal this lump as a nest, so perfectly blended onto the branch that it would not have been recognised, but for the bird's close attention.

With the nest almost finished, the fantail brings cobwebs and smears the sticky web threads around the outer surface to make a firm, felt surface, its grey matching the colour of the branch upon which it is built.

This neat little nest, its bowl scarcely larger than that of a kitchen eggcup, merges not only to the supporting twig, but tapers smoothly to a long fine tail. It is almost the same as the nest of the much more common and widespread Grey Fantail, and both are just smaller versions of the Willie Wagtail's nest (except that the Willie Wagtail's nest has no tail).

These southern Rufous Fantails are migratory, moving north from March to April to spend the winter in north-eastern Queensland and southern parts of Papua New Guinea. Around September to October they return to the south-east (including the Greater Brisbane area) to breed during the summer months. Rufous Fantails of far northern Queensland tend to be only locally nomadic through winter because their region is warm enough throughout the rest of the year.

The Rufous Fantail has been recorded through many of the Greater Brisbane area's suburbs, but mostly in outer areas where larger lots provide more vegetation; and especially near rainforest, dense forest or parkland reserves.

name: Rufous Fantail *Rhipidura rufifrons*

length: 15–16 cm

looks: unique, small very active bird; very long and colourful tail, which most of the time is kept widely fanned; tail bright orange-rufous towards its base, and is almost constantly swung side to side

voice: high, weak, squeaked "tsit" and "tsi-tsit"; song is a tinkling sequence of these sounds

habitat: rainforest, dense wet eucalypt and monsoon forests; paperbark and mangrove swamps, open country on migration

nesting: neat, tiny cup nest with long tail, similar to that built by the more common and widespread Grey Fantail

abundance: quite common in suitable habitat in SE Qld

Lewin's Honeyeater *Meliphaga lewinii*

name: Lewin's Honeyeater *Meliphaga lewinii*
length: 19–22 cm
looks: large, honeyeater with dark plumage; bright yellow, semicircular ear patch; if seen close at hand, a fine yellow line from corner of bill to back under the eye; yellow tint to wing feather edges
voice: very loud, far-carrying knocking or (when fast) rattle; varied from slow to very fast; also known as "Machine-gun Bird"
habitat: rainforests, wet eucalypt forests, woodlands, heathlands
nesting: rather bulky cup nest, hung by rim between twigs in dense foliage at 2–5 m high
abundance: abundant

The majority of Australian honeyeaters are inhabitants of woodlands, heathlands and open forests, where flowering trees and scrubs are most bountiful. The Lewin's Honeyeater is one of very few that is commonly found in rainforest. In south-eastern Queensland it also occurs widely in those other vegetation areas.

The presence of this large, dark honeyeater is often revealed by its call — a loud, machine-gun rattle, slow to very rapid. In rainforest the call seems to reverberate through the cavernous spaces beneath the canopy, and carrying far from where the bird is calling.

Individual birds or pairs hold a territory most of the year, advertising ownership with their calls; usually it is only during winter that some wander beyond territory boundaries and into other habitats.

This is an aggressive species, spending much time chasing its own kind and other species of honeyeater away from most favoured feeding places. When they visit gardens, these birds often try to dominate any natural or provided food source.

In rainforests the Lewin's Honeyeater feeds at mid to upper levels, in pairs or alone. At times this bird hovers to take nectar and insects from the flowers of trees; at other times it takes insects from the foliage or climbs about tree limbs probing insects and spiders from bark crevices. Fruits are taken from rainforest vegetation, and at times from gardens and orchards.

The plumage of this large honeyeater blends well with the universal greenery of the rainforests, being mostly dark to pale olive green. Only a pale yellow streak back from the corner of the mouth to just below the eye, and a large, conspicuous pale yellow, crescent-shaped ear-patch, relieve what is otherwise a very gloomy plumage. Male and female are alike.

In south-eastern Queensland the breeding season is from spring to early summer. The nest is suspended by its rim in the dense foliage of a rainforest tree or tall shrub at any height from 2–8 m. It is a strong but rather untidy cup of bark strips, leaves and moss, bound with webs, and lined with soft plant down. The two eggs are white, spotted and blotched with dark red.

Scarlet Honeyeater *Myzomela sanguinolenta*

Their almost constant tinkling calls, like tiny silver bells in the treetops, may draw attention to the leaf-sized Scarlet Honeyeater. More often heard than seen, they are the smallest of Australian honeyeaters.

While mostly feeding in the upper parts of forest and woodland trees, these honeyeaters do occasionally descend to lower flowering shrubs (banksias, grevilleas and bottlebrushes) in heathlands and suburban Brisbane gardens. Scarlet Honeyeaters feed in loosely scattered groups, often mingling with other small honeyeaters through the treetop canopy or shrubbery where there are plenty of flowers. They move quickly to different flowering species, each offering a new nectar source. For a few days there may be many Scarlet Honeyeaters in a garden before being absent for months.

When low in a garden's shrubbery, their calls can hardly be missed. But in the tall forest trees, their pleasant and almost continuous calls carry only so far, so when calls come from many of these birds, they reveal the presence of a species that, otherwise, almost certainly would have been missed.

The singing seems to be mostly from the males. Their calls mark their feeding territories, but are also used during the time of nest-building. The call is a rather piercing, high "tsieep, tsip, tsieep", while the male's song is a brisk and cheery, erratically rising and falling "chwip, swit-sweet-switty-swit-switty".

If the calling birds are sighted, high in treetop foliage, binoculars will be needed to appreciate the intense plumage colours of the male — his head and body entirely scarlet, the wings a very dark brown. Occasionally, when low in shrubbery (perhaps even in a native garden), the beauty of these tiny birds can be observed up close.

This honeyeater makes use of a great variety of habitats — edges of rainforests, eucalypt forests and woodlands; paperbark swamps; heathlands; acacia scrublands; and suburban parks and gardens.

Breeding is around August to January, and there are often two broods. The female builds a small cup nest of fine bark strips, rootlets and grass, bound with cobwebs and suspended by its rim from green, leafy twigs of the outer foliage at heights of 1–10 m.

name: Scarlet Honeyeater *Myzomela sanguinolenta*

length: 10–11 cm

looks: male unique, with head and body entirely scarlet, wings and tail black; female mid grey-brown above, off-white beneath, faint pink on chin and throat

voice: call is a sharp, short, whistled "tseeip"; song is a silvery, tinkling tune rising and falling

habitat: rainforests (especially their margins), wet eucalypt forests, dry eucalypt forests, woodlands, paperbark swamps, gardens with rainforest trees or densely planted with flowering trees and shrubs; feeds on nectar and insects

nesting: small, delicate cup nest of fine strips of bark and grass, bound with webs; suspended by the rim from thin twigs of the outer foliage

abundance: common, but erratic visitor to some parts of its range

Logrunner *Orthonyx temminckii*

name: Logrunner *Orthonyx temminckii*
length: 18–21 cm
looks: colourful small bird of the rainforest floor; plumage of both sexes quite bright cinnamon-rufous, patterned and streaked with dark brown, black and white; underbody of male (from chin to legs) is all-white; female is more colourful with bright orange chin to breast, then white back to legs
voice: noisy at dawn and dusk, with loud racket of clattering "kwik-it , quikit", "queeek-queeek" and "be-queek, be-queek"
habitat: rainforests, usually at higher altitudes
nesting: nest usually on ground or on a stump; nest built by female; platform of sticks, then walls, finally roofed over, covered with moss and leaves, difficult to detect; remains quite dry in rain
abundance: moderately common in a rather broken and restricted habitat

The sharp, loud clatter of Logrunners (as several birds do battle with song) is likely to dominate parts of the rainforest at twilight. Several birds calling to, or at, each other is heard most often at dawn and dusk.

This is probably the easiest way to find these birds. Keep an ear open for the loud, sharp, vigorous and rapidly repeated "be-queek, be-queek". For the rest of the day they forage on the ground, giving only an occasional quick sharp contact call or soft chatter, and are not easily sighted in that cluttered shadowy habitat. Logrunner calls are, however, more likely to be heard from April to July, early in their breeding season, when males may call frequently throughout the day.

This bird has also been known as the Spine-tailed Logrunner, due to the strong spines protruding from the tail tip. The tail tip is often used as a support as they stand on one leg and scratch vigorously into the leaf litter and soil surface. Although these birds are colourful, they merge with the colours of the rainforest floor, matching well the orange and grey of fallen leaves, the browns and black of decayed logs, rocks and earth.

The Logrunner is a common bird of the rainforests of the high ranges that almost encircle the Greater Brisbane metropolitan area. It would once have been found in places now occupied by city and suburbs but, with rare exception, the sightings of this species now come from the outermost suburbs near or on forested ranges.

Many of those "backyards" are large areas, enough to retain some wet forest, and perhaps adjoin a larger area of forest or national park. For those fortunate enough to have a property that is home to Logrunners, or close to known Logrunner territory, it may be possible to conserve, increase or restore areas of rainforest or dense wet eucalypt forest, or similar vegetation.

Unfortunately, the owners of Brisbane's small, inner-suburb blocks are unlikely to ever have Logrunners in their own backyards. This would be a species to seek out on a day trip to one of Brisbane's nearby national parks. In some popular parks, walking tracks with a steady flow of visitors have made these normally timid birds less wary. The loud calls provide the best opportunity for a sighting; otherwise, listen for small movements and rustlings among the leaf litter.

Black-faced Monarch *Monarcha melanopsis*

The Black-faced Monarch is quite a large flycatcher, readily identified by its black mask, which is conspicuous against the silvery light-grey of the rest of the head. The black surrounds the bill and extends down onto the throat, but it is the beautiful rufous tan of the underparts that makes this species especially attractive.

This is a bird of coastal rainforest scrub, heavy wet eucalypt forests and damp fern gullies, but it frequently ventures out of these haunts into more lightly timbered country. The monarch feeds among the forest's mid-level foliage and branches. It moves rather slowly, sedately, occasionally picking some small creature from a leaf while fluttering about the foliage, but only rarely taking insects in flight.

The voice of the Black-faced Monarch is distinctive enough that, once known, it can assist in finding these birds in their usual dense (and often rather gloomy) habitat. The usual call is a series of deliberate whistles, which rise then fall — "wheech-iew, wheech-iew, whieeeuw". Also in its repertoire, as a single call, or at widely spaced intervals, is a "wheeit-chiew", and a very husky, scratchy "skreeeit".

In the southern parts of its range, including the Greater Brisbane area, it arrives from northern Queensland in early September, with breeding activity beginning in October. Both sexes share the nest-building, choosing a site as low as 1 m or as high as 10 m. A fork of a shrub or small tree is usual, and an open and exposed aspect is preferred.

The nest is an open cup, with thick walls of mosses and plant fibres, bound with webs and covered with a thick outermost layer of green mosses. The nest resembles a clump of mossy growth on the branch and is therefore able to survive in the exposed position in which it is usually placed. The two white eggs are spotted with red.

This is a species that could visit Brisbane gardens — more likely the larger lots, where there is some remnant rainforest or other rather dense vegetation, or homes close to a park or forest reserve containing suitable habitat. Most likely these birds would be just passing through, perhaps on their annual migratory travels, rather than staying and nesting. But depending upon the locality and suitability of the "backyard", it is always a possibility they could be in for a longer stay.

name: Black-faced Monarch *Monarcha melanopsis*

length: 16–20 cm

looks: head, back and wings pale grey; breast grey; rest of underbody rufous-buff; forehead to throat black; immature similar, but without black face

voice: mellow whistle, each time rises, then falls "wheeit-chiew, wheeit-chiew"

habitat: rainforests, mangroves, eucalypt forests and woodlands

nesting: moves in spring to secluded gullies where vegetation is dense; nest built in fork of understorey sapling, 2–5 m high; rounded bowl of long, needle-like casuarina "leaves", intermixed with green moss, giving the exterior a green mossy appearance; although typically on a bare twig, without screening foliage, nest is well camouflaged

abundance: quite common in optimum habitat

Noisy Pitta *Pitta versicolor*

name: Noisy Pitta *Pitta versicolor*

length: 18–20 cm

looks: large colourful pitta, more easily identified as it is the only pitta to occur in SE Qld; shape and behaviour of bird may be more useful than colours in dim rainforest lighting; short-tailed, rather big-headed shape; scratches about in litter of rainforest floor for snails and other food

voice: call commonly described by words "walk-to-work"; easily remembered once heard, and carries far through the rainforest

habitat: rainforests, wet eucalypt forests

nesting: nest built between buttress roots of rainforest tree, or similar enclosed space; nest is quite a massive collection of thick sticks built up as a domed spherical nest with side entry; accumulated moss and leaves on top make it difficult to recognise

abundance: quite common in its restricted extent of habitat

So often heard through spring and summer months, the loud calls of the Noisy Pitta must be one of the best-known, most characteristic sounds of the rainforests in the ranges around Brisbane. But the bird itself, although brightly coloured, is usually shy, elusive and difficult to see in the dim light and amongst the vegetation and debris of the forest floor.

Pittas are rather stumpy, ground-foraging birds of south-east Asian rainforests and other parts of the Northern Hemisphere. Four species occur in northern Australia, but only one, the Noisy Pitta, graces the rainforests around Brisbane.

When seen, the Noisy Pitta is larger than depicted in illustrations (where its compact shape is not unlike various rather plump little birds that scratch about or hunt on the forest floor, particularly the scrubwrens). But the Noisy Pitta is 20 cm in length — mostly head-and-body for its stump tail barely reaches beyond its folded wings. Common birds of roughly similar body length would be the Sacred Kingfisher and the Magpie-lark, so this pitta is not a small bird.

It is the brightly coloured plumage of the Noisy Pitta that catches the attention when one is sighted. The pittas are such a colourful family, they have been known as "jewel-thrushes". Brisbane's Noisy Pittas are no exception, plumed in green, yellow, blue and red. While these colours often do not show up well in their gloomy rainforest habitat, should one of them be exposed by sunlight, with the light playing on the plumage, the true beauty of it will quickly be revealed.

This pitta fossicks about on the forest floor, searching among the leaf litter for worms, insects, wood-lice and, a particular favourite, very large rainforest snails, the shell of which is broken against a rock. This habit has led to an alternative name, "Anvil Bird", which is now rarely used for this species.

Throughout their spring and summer breeding season the Noisy Pitta calls loudly and often — a sound commonly put into words as "walk to work". The nest is a large, rough pile of sticks, with a side entry, tucked between the buttress roots of a tree or other protected location.

Paradise Riflebird *Ptiloris paradiseus*

Even when calling loudly, the Paradise Riflebird may not easily be detected. It seems to keep mostly to the higher limbs of tall trees, moving in a rather treecreeper-like manner over bark and dead wood. The long, strongly curved bill is used to search out small prey. It prises into crevices and sifts through debris caught in the ferns and orchids that festoon those high limbs. At times (when not too high up), its position may be revealed by the sounds of its rustling and tearing at bark or rotting wood.

Females and immature birds seem to be sighted more often than males. Perhaps less wary and more inclined to descend lower to mid-level vegetation, their rich cinnamon-rufous plumage often makes them more conspicuous than the deep, blue-black males.

Each mature male Paradise Riflebird has display perches in his territory. Typically they are high horizontal limbs of tall rainforest trees, but may occasionally be quite low. These sites he visits at intervals through the day, to call, and if there is a female in the vicinity, he will launch into flamboyant display.

Wings are fanned vigorously up and down, producing a loud rustling sound and clapping sounds. The head is thrown far back and moved rhythmically from side to side. The bill is opened, showing the bright yellow mouth. These movements display to full effect the rich iridescent colours of the bib-like panel from throat to breast. With every movement there are changes in colour, from black with a violet gloss, to shimmering turquoise blue then, with another change of posture, transforming to a vivid purplish-magenta sheen.

The calls include a harsh, rasping "yaaarsss" — a series of sounds increasing in strength and volume, then fading away to a soft hiss. At times it also gives a long upwards whistle and soft churring sounds.

The Paradise Riflebird occurs along the southern edge of the Greater Brisbane region; in the McPherson Ranges; and in the Bunya Mountains to the north. While temperate and subtropical rainforests are its primary habitat, the species forages into the tree foliage of nearby wet eucalypt forest and paperbark swamplands. This is the only riflebird in southern Queensland; two others occur in the north-east and Cape York Peninsula.

name: Paradise Riflebird *Ptiloris paradiseus*

length: 28–33 cm

looks: only riflebird in SE Qld; males black with a glossy iridescence of green, blue and violet; bill long and strongly downcurved; when looking beneath this bird from a distance, the small differences between it and far-off NE Qld riflebirds would be difficult to distinguish

voice: slow, drawn-out, rasping sound, louder, then fades to a hiss, as in "yaaarsss"

habitat: subtropical and temperate rainforest

nesting: nest built by female; bulky bowl of rootlets, vine tendrils and leaves, built into a dense clump of foliage

abundance: moderately common in small area of habitat

Pale-yellow Robin *Tregellasia capito*

name: Pale-yellow Robin *Tregellasia capito*
length: 12–13.5 cm
looks: small robin; head grey-brown with large white forehead patch joining a white throat; back and wings olive; breast to undertail covert pale yellow
voice: metallic squeaks "chweeik, chweeeik", and scolding "scraich"
habitat: rainforests, preferring areas of dense undergrowth; occasionally into adjoining wet eucalypt forest
nesting: neat, deep cup, camouflaged with moss and bark; often built onto spiny stem of a lawyer vine
abundance: common in rainforest habitat

This is a small, unobtrusive and rather tame robin of the rainforests. It is easily overlooked in its gloomy home, for not only is this small robin found only in rainforest, it prefers the most dense areas of undergrowth along creeks and in thickets of tangled lawyer vine.

The behaviour of the Pale-yellow Robin also makes it less conspicuous. It forages in the mid to low levels of the rainforest, perching still and silent on low branches and intently watching, occasionally dropping to the ground to take some very small creature from the leaf litter, a mossy log or the damp earth.

Compared with the well-known, larger Eastern Yellow Robin, it is much less active. When perched, it sits for long periods silent and motionless, with little of the tail-jerking, wing-flicking restless movements typical of that bird. It is less likely to reveal its position by flying often from perch to perch, and has a reputation of seeming tame, or confiding — perhaps because of this tendency to sit still and seemingly be unconcerned at the presence of an observer.

While the Pale-yellow Robin resembles the larger Eastern Yellow Robin in shape and colour, and in its habit of clinging sideways to a trunk or vine stem while scanning the ground, in other ways it behaves more like the smaller flycatchers. At times it flutters about the mid-level foliage of the rainforest, taking insects in flight. Its bill is more like those of small flycatchers — a rather flattened, wider shape.

The Pale-yellow Robin breeds from August to December, building a small, neat cup into a slender fork of a rainforest shrub, or higher in a small tree, at heights of 1–3 m. The nest is often built onto a lawyer vine — its sharp needle points and hook-edged tendrils offering considerable protection. The exterior is camouflaged with a layer of green, threadlike mosses, and decorated with flakes of pale bark that make it look like part of the surrounding tangled moss and fern-covered debris. The female alone builds the nest and incubates; the male, however, does feed her on the nest.

Throughout the year these birds keep to their territory of a few hectares, and for most of that time are solitary. Sometimes they stay with the young of previous nestlings.

Rufous Scrub-bird *Atrichornis rufescens*

The Rufous Scrub-bird is famed for its powerful and varied song. This peaks early in the spring breeding season when the loud, ringing and penetrating "cheip-cheip-cheip" calls can be heard from many hundreds of metres away. Despite a male's calls becoming almost deafeningly loud, he is rarely seen, keeping to the most impenetrable parts of the forest. The female is usually silent and even more difficult to see.

Mimicry of other forest birds greatly widens the scrub-bird's repertoire of song. It is said to include calls of the Logrunner, Golden Whistler, Rufous Fantail, various scrubwrens and other forest birds. These snatches of mimicry seem to be included also in their quieter songs outside the breeding season.

The Rufous Scrub-bird is quite rare. Within south-east Queensland it is restricted to high range tops (the McPherson Range) where most of its populations are within national parks. Before about 1900, this species was an inhabitant of lowland rainforests, but that habitat has almost entirely been cleared.

The Rufous Scrub-bird inhabits temperate and sub-tropical rainforest, and nearby wet eucalypt forest. It usually lives in the dense, ground-covering undergrowth that develops beneath gaps in the forest canopy or perhaps where a tree has fallen. There it scratches about in the leaf litter and exposed, moist surface soil in search of insects, land crustaceans, worms and snails.

It is thought that this species, and Australia's two species of lyrebird, may have had their origins in cool damp Antarctic beech forests that were once more widespread on the continent. Now they seem to keep to the shady and moist, coolest parts, and breed in the coldest months. Their nests are well suited to these cold, wet situations — the scrub-birds and lyrebirds build substantial, domed and thickly lined nests.

The nest of the Rufous Scrub-bird is only rarely found, probably because the female (that alone builds the nest, incubates the eggs and tends the nestlings) is so secretive and far more difficult to follow than the noisy male. The nest is always extremely well hidden in a clump of sedge, mat-rush or pile of debris. The exterior is camouflaged with dead leaves, pieces of ferns and twigs; the interior is lined with a layer of dried wood pulp obtained from decaying logs on the forest floor.

name: Rufous Scrub-bird *Atrichornis rufescens*

length: 17–19 cm

looks: overall rufous, with dark brown barring across wings and long tail, and lightly on head; lighter cinnamon-rufous of underparts darkly barred on throat and sides of neck; bold white streak across face

voice: powerful ringing calls; "cheip-cheip" can be heard from afar and almost painfully loud when near

habitat: temperate rainforest at altitude about 600–1000 m; previously in lower altitude rainforests now cleared

nesting: nest built and attended by female, usually under a dense clump of rushes of the rainforest floor; large nest for size of bird; a round thickwalled sphere woven of blady grass, rushes and ferns, and lined with a cardboard-like layer of dried softwood pulp

abundance: fragmented population on high range tops

Wonga Pigeon *Leucosarcia melanoleuca*

name: Wonga Pigeon *Leucosarcia melanoleuca*
length: 35–40 cm
looks: large, ground-foraging grey-and-white pigeon; upperparts, neck, wings and tail dark grey, white heavily flecked dark grey. Distinctive double 'V' shape on front of neck and breast.
voice: long series of monotonously identical calls, 'whoik-whoik-whoik-' all at same pitch and frequency
habitat: rainforest, thickets, tall wet eucalypt forests
nesting: builds nest up to 20 m above ground in substantial fork or vine tangle; or much lower on stump or treefern crown
abundance: generally uncommon, but locally quite prominent in areas of optimum habitat

In tall wet eucalypt forests and rainforests probably the most continuous, persistent and, after a time, most monotonous of call is that of the Wonga Pigeon. The high clear "whoik-whoik-whoik" goes on and on, for long periods, without variation or pause.

In some places where food is abundant under certain trees (those dropping seeds and ripe fruits), there may be a few of these pigeons feeding together; however, most of the time the Wonga Pigeon walks and fossicks about the forest floor alone. This feeding routine tends to occur more in the early morning and late in the day; through the middle of the day this bird settles into its routine of calling from the ground or a low perch.

The nest is typically a sparse and flimsy construction, quite large at some 30 cm in diameter. It is positioned in a vine tangle, in the crown of a tree fern or among the branches of a tree, at a height of 3–20 m.

Emerald Dove *Chalcophaps indica*

name: Emerald Dove *Chalcophaps indica*
length: 23–27 cm
looks: light pinkish-brown except wings, which are rather iridescent green; large flight feathers and tail are a deep rufous-brown; underwings similar to body colour
voice: advertising call a low purring-moaning; repeated a considerable many times, each time stronger, higher
habitat: rainforests, monsoon forests, wet eucalypt forests, melaleuca woodlands, lantana thickets, regrowth scrub
nesting: rough, usually rather flimsy stick nest, hidden in dense foliage
abundance: moderately common

In the gloomy lighting of its dense forest home, where the Emerald Dove forages on the ground, its colours appear quite dull. But seen walking on the sunlit ground of a forest clearing, the brilliant iridescent green of the plumage is indeed appropriately described as emerald.

The green of the wings are also in strong colour contrast against the violet-brown of its body and the chestnut of its outer flight feathers, although the latter is only likely to be glimpsed as the dove takes flight.

This is a rather solitary bird, moving slowly about the ground under dense forest or regrowth, seeking fallen seeds; sometimes it will walk along branches to take rainforest fruits. The call is a series of low moaning sounds that each time rise higher, repeated some five to ten times before the bird pauses to begin another series.

Typical of most doves, the nest is a very rough platform of sticks, slightly dished downwards at the centre and built into a bush, tree fern, or into a tangled mass of vines.

Eastern Whipbird *Psophodes olivaceus*

The name "whipbird" comes from the loud, ringing whip-crack calls of this long-tailed, predominantly olive-green bird. But despite its far-carrying calls, it can be very difficult to see, keeping to thickets and densely tangled undergrowth. The calls of the Eastern Whipbird are one of the most common and characteristic sounds of the rainforests and densely vegetated outer areas of Brisbane and its encircling forested ranges.

The call is often given as a duet between male and female, with the female's answering call so perfectly timed to follow that it might be thought that this is but the call of one bird. However, when on some occasions the female does not answer, the difference is obvious. The male's call is a drawn-out high whistle, beginning very softly, building in volume to a powerful whip-crack, an almost explosive sound ringing through the forest, and heard from afar. The female usually answers immediately, a sharp but much less noisy "tchew-tchew".

This species has adapted and learned to live in areas of human habitation. It is commonly heard calling from patches of "acreage" backyards — those properties with, or adjoining, dense thicket vegetation. It occurs in some suburbs that are not far out from the city centre. Chapel Hill, with lots of areas up to half a hectare or more, has its resident Eastern Whipbirds, and they are widespread through many, if not most, outer districts.

This species has been able to move from its natural dense vegetational habitat to use backyard, creekside, roadside and vacant land vegetation that has a similar dense rainforest character, but often not many native plant species: foremost of these are lantana thickets.

Owners of large, and even quite small backyards in some suburbs, may be able to encourage Eastern Whipbirds, along with the many other birds that need a dense and concealing vegetation. Planting dense native shrubs, perhaps creepers, and beneath them, tree ferns and smaller ferns will work nicely. The ground needs to keep its cover of damp and decaying fallen leaves, beneath the overhead foliage cover. Such a garden (or more practically a part of the yard), even if somewhat untidy to some human eyes, will be an invitation to the whipbird and many others species.

name: Eastern Whipbird *Psophodes olivaceus*
length: 25–30 cm
looks: mostly olive green; crested black head and breast; bold white cheek streak
voice: drawn-out, rising whistle ending in explosive whip-crack, usually answered by sharp "tchew-tchew" from female
habitat: margins of rainforests, tickets, forests, heaths
nesting: cup of fine twigs and bark, 0.5–2 m above ground, 2–3 eggs; July–Dec
abundance: common

Urban Birds

Brisbane today, although largely an artificial rather than a natural environment, continues to enjoy the advantages of its location in a region of rich and varied plant and animal life.

Within a few kilometres of the city centre, Scarlet Honeyeaters feed among crimson bottlebrush flowers at the foothills of Mt Coot-tha, while Azure Kingfishers and Striated Pardalotes dig their nest tunnels into creekbanks. Rainbow and Scaly-breasted Lorikeets noisily raid the nectar of jacarandas and silky oaks in streets and gardens. In tall grass along roadsides and vacant blocks of land, cisticolas stitch together their nests of leaves, and Red-backed Fairy-wrens skilfully conceal their small domed nests. Across the treetops of outer suburbs (and some inner suburbs), the Rainbow Bee-eaters, Crimson Rosellas, Dollarbirds and Figbirds are conspicuous. Less obvious are the many other birds, large and small, which inhabit the gardens and parks of the suburbs, from large Australian Brush-turkeys, Pheasant Coucals and Satin Bowerbirds to the various small honeyeaters doing their rounds among flowering shrubs.

Few other parts of Australia can compare with the diversity of bird species found within the Greater Brisbane area. No doubt this is enhanced by the diversity of habitat. In less than two hours by car there are many natural ecosystems — islands, coastal waters, mangrove swamps, heathlands, mountain ranges (with both the subtropical and temperate rainforests), wet eucalypt forests, open forests and woodlands. The rich bird population of the suburbs, especially the large blocks of outer suburbs, could be expected to have many birds. But the very small lots of inner suburbs also support a surprising variety of bird residents and visitors. It is in these suburban backyards that the birds of Brisbane can be encouraged not only to remain, but to increase in variety and number.

Gardens, streets and parks can be planted out with some of the nectar-producing, flowering trees, shrubs and small ground cover plants, that will attract many species of nectar-seeking birds occurring throughout south-eastern Queensland. Where there is room for larger trees, as well as nectar, they may provide berries or fruits, attracting various fruit-eating dove and pigeon visitors.

Some parts of gardens with room, perhaps at the back of the yard, may be left a little overgrown, or create a space for some dense shrubs with leaf litter on the ground. Small ground-fossicking birds, like scrub-wrens, can scratch about and perhaps conceal their nests in there. Even so simple a contribution as a bowl set under a tap, can encourage birds to a garden. Brisbane and its surrounds are blessed by an abundance of natural attractions, and with just a little effort from homeowners, habitats for the region's native birds can be increased.

TOP BIRDWATCHING sites

BRISBANE FOREST PARK

This large bushland area lies only 12 km from Brisbane's city centre. It is a magnificent reserve of some 28,500 ha, encompassing most of the high ranges around Mt Glorious and creating a magnificent green backdrop to the city. Brisbane Forest Park protects over 800 plant species and 200 types of native animal (including numerous birds).

BAYSIDE PARKLANDS

A series of small parklands spread over 16 km along the shores of Moreton Bay have a combined area of over 500 ha. Mangroves, mudflats, tidal creeks, salt marshes wallum heathlands and open forests lie just 15 km to the east of Brisbane CBD, with access via Wynnum Road, Manly Road and Chelsea Road.

Collared (Mangrove) Kingfishers and Mangrove Gerygones are usually present along the Mangrove Boardwalk at Wynnum North. To find the Wader Roost, take the path opposite the entrance to the car park. This is best used at high tide when, among other waders, the Asian Dowitcher has been seen.

Left: Laughing Kookaburra.

Australian Brush-turkey *Alectura lathami*

name: Australian Brush-turkey *Alectura lathami*
length: 60–70 cm
looks: very large, black, turkey-like bird; unique vertical tail; bare red neck
voice: low vibrating grunts
habitat: rainforests, wet eucalypt forests
nesting: megapode, which builds incubator mound, where warmth generated by fermenting vegetation hatches eggs
abundance: common

Found throughout the Greater Brisbane region, and no doubt familiar to almost all residents, this bird is so large and distinctive in appearance that it is hardly likely to go unseen. Most conspicuous are the bald red head and neck, and the unique vertical fan-shaped tail.

Although usually seen on the ground, this heavy bird can fly and climb, and roosts quite high in trees. It is extremely alert and timid in the wild, but becomes tame in tourist parks, and often also in suburban gardens. Its presence may be revealed at times by the calls. The males give deep booming and grunting sounds, while both sexes give various softer low groans and grunts.

The Brush Turkey is typically a permanent resident, with restricted local wanderings, but probably expands its home territory in the breeding season. Young birds wander further before settling into a territory of their own.

No true nest is built; these birds are megapodes — laying eggs into a huge pile of leaves, twigs and earth. Like compost, the decaying vegetation generates enough warmth to incubate the eggs buried in the centre of the mound. Mounds are used for many years. Each year around May, if there has been sufficient rain, the male rebuilds the mound, scraping in what new leaf and twig debris he can find.

Egg-laying and temperature maintenance continue for some 35–45 weeks, with a female laying perhaps 25 eggs in the mound over that period. Upon hatching, the young struggle up to the surface and run off, completely independent of their parents.

The Australian Brush-turkey is more likely to occur in the larger, more heavily vegetated backyards of the outer suburbs, but may visit smaller yards that offer suitably dense vegetation. Suitable suburban habitat can be found where there is probably some remnants of rainforest or wet eucalypt forest, and lower understorey trees, shrubs, or other dense garden shrubbery. In such situations, across a number of adjoining backyards, there could be found a suitably large area of habitat that enables these adaptable birds to utilise parts of the most built-up suburbs for their habitation.

Pheasant Coucal *Centropus phasianinus*

This large, beautifully patterned bird is a member of the cuckoo family, but one which has gone off in its own direction to be quite unlike any other cuckoo in Australia.

Other cuckoos are birds of the trees — strong fliers that are well known for avoiding the hard slog of parenthood by laying their eggs in nests of other bird species. But the Pheasant Coucal is a ground-dweller — weak and clumsy in flight, and committed to building its own nest and raising its own young.

Pheasant Coucals live and hunt beneath the cover of dense low vegetation, especially tall, rank, dense grass, lantana thickets, swampy heathlands and mangrove fringes. It is often a resident in vacant suburban lots and margins of creeks and drains that are overgrown with tall grass, lantana or other weedy vegetation.

Out of sight beneath this dense concealing cover, the Pheasant Coucal runs, almost reptile-like, along pathways and tunnels, hunting frogs, small reptiles and mice. It also raids nests of small birds.

Often the only indication of its presence is its call — a long series of rising and falling "coop, coop, cook-cook-cook-cook". Another call is a rather metallic, tapping sound "chak-chak-chok, chok, chowk, chowgk". Birds are often accompanied by a second bird calling at a slightly different pitch.

Mostly the Pheasant Coucal stays out of sight. But it will, when its plumage is wet from rain or sodden vegetation, climb up onto a higher branch or post. With plumage spread to dry the very long tail may be seen — nearly half of this bird's length.

Occasionally the Pheasant Coucal will clamber much higher in a tree. Then when returning to its ground-level vegetation, it descends in a steep glide, on short rounded wings, to plunge from sight into the grass. Despite their reluctance to fly, Pheasant Coucals do at times travel high and for quite long distances.

The Pheasant Coucal also differs from other Australian cuckoos in building its own nest, which is a large shallow bowl of trampled leaves of reeds or grass, well hidden on a clump of dense vegetation and usually within half a metre of the ground.

name: Pheasant Coucal *Centropus phasianinus*

length: 60–75 cm

looks: superficially pheasant-like member of the cuckoo family; very large; mostly terrestrial; rufous plumage heavily streaked and barred, especially the very long tail; male has white-streaked black head

voice: long, loud, resonant series of notes "coop-coop-cook-cook-cook"

habitat: tall, rank grass of waste lands, river and swamp surrounds, swampy heathlands, lush grass and overgrown urban lots, roadsides, drains

nesting: rough trampled platform of reeds, grass, cane; well hidden in thickest vegetation

abundance: common

White-throated Gerygone *Gerygone olivacea*

name: White-throated Gerygone *Gerygone olivacea*
length: 10–11 cm
looks: tiny warbler; light brownish upper parts; pure white throat; light bright lemon-yellow breast
voice: strong, clear, far-carrying, a sequence beginning high and seeming to tumble or cascade downwards
habitat: open forests, woodlands, parks and gardens
nesting: small neat domed nest with side entrance, suspended by roof from twig of outer foliage; 6–12 m high
abundance: common

With a wide range, which includes Brisbane and most other cities and major towns of eastern Australia, this is one of the best known gerygones. Little more than leaf-sized, these small birds would likely remain almost unseen but for their exquisite song carrying far through woodlands, parks or gardens.

The song usually begins with a few piercing notes, these followed immediately by a long, clear undulating sequence of notes descending, briefly rising, then lowering again, and repeated with minor variations.

These birds forage for insects through the outer foliage of all kinds of trees. For a treetop bird, areas of lawn with trees, neglected paddocks of tall grass, or almost any other ground cover would be a fine substitute for natural open forest or woodland.

This species builds a domed nest with a side entrance. It is suspended from its roof by a thin cable of grass and webs attached to a slender twig of the outer foliage of a tree. There is a long tapered tail hanging from the nest.

Little Wattlebird *Anthochaera chrysoptera*

name: Little Wattlebird *Anthochaera chrysoptera*
length: 32–36 cm
looks: large honeyeater; dark plumage densely streaked with fine white markings; rusty tone on wings is more obvious in flight, as are white tips on wings and tail
voice: varied, from rather musical, abrupt "kook" to grating "kaark"
habitat: forests, woodlands, wallum heathlands, parks and gardens; feeds on nectar, insects
nesting: rather untidy bowl of sticks, bark, lined inside with finer materials
abundance: common, nomadic

The Little Wattlebird must be one of the more familiar of the birds of urban gardens. It makes its presence obvious, being noisy and at times aggressive towards other birds, particularly the smaller honeyeaters.

The call of the Little Wattlebird rather suits its changeable and at times abrasive personality. There is often a mellow "kook" and ""koo-op", often intermixed with a grating sound, "kook-kook-kraagk" usually repeated several times.

The plumage is overall grey-black with fine white streaks. There is rufous on wing feathers, often not visible on folded wing, but in flight making a rufous panel along each wing. Also conspicuous in flight are the white tips to the wings and tail feather tips.

The Little Wattlebird breeds almost any time of the year, however the months of July to December would be usual. The nest is a large cup of twigs and bark, well lined with soft grass and feathers, and placed in a vertical fork of tree or shrub at height between 3–20 m.

Golden-headed Cisticola *Cisticola exilis*

This tiny bird inhabits the lush tall grasses and other dense low vegetation of swampy ground, where it skulks low under cover and is seldom seen or heard. But there comes a transformation in the breeding season, through spring and summer, when the male calls conspicuously and almost incessantly. At this time, out in the open, his golden plumage, and especially the deep golden-orange of the crown of his head, can be admired.

The male's buzzing carries far, and could be mistaken as insect noise. In contrast to his secretive behaviour at other times, he calls from an exposed perch on the top of tall grass clumps, branches, fences or overhead wires. In his fluttering song-flight across his nest territory, undulating or bouncing, the male Golden-headed Cisticola keeps perhaps 10–20 m above the ground, all the while pouring forth drawn-out, metallic-buzzing sounds, often interspersed with clear musical "teewip" calls.

These birds are common though the suburbs wherever there is vacant land, perhaps part of a large backyard that holds shallow pools surrounded by tall grass or weeds. The lush edges of roadside drains, flooded river flats, the samphire margins of coastal wetlands, irrigated pastures and channels are also suitable habitats for the Golden-headed Cisticola.

The Golden-headed Cisticola is also known as the "Tailor-bird" for its distinctive nest, which is enclosed in large leaves that have been pulled in close around the nest with their edges "stitched" together.

The nest itself is a rounded dome with an entrance at the side, towards the top. Its neat, softly textured walls are made of fine grass and plant down, bound with spider webs and decorated with their silken cocoons. It is usually less than 1 m above ground and well concealed in dense plant growth. Where possible, large leaves from the surrounding vegetation are pulled close against the nest and together stitched to it, thereby effectively concealing it from view.

Male and female work together to stitch the leaves, the male staying outside the nest to pass through strands of coarse spider web and fine fibres to the female inside. She attaches the webs; they stick and hold the large soft leaves against and around the nest. Although the male helps in the construction of the nest, the incubation is done entirely by the female.

name: Golden-headed Cisticola *Cisticola exilis*
length: 9–11 cm
looks: small; overall golden; deeper orange on head; dark brown streaking on back, wings and tail; underparts light golden-buff; female and non-breeding male similar but much paler
voice: in spring and summer calls almost continuously; a metallic buzzing sound interspersed with clear "teewip" calls
habitat: damp, tall grass margins of swamps, tall grass in damp pastures
nesting: small domed nest, to which are bent down and attached green leaves of bush in which nest is built; tall grasses of vacant suburban lots, roadsides, swampy pastures of suburbs
abundance: common

Superb Fairy-wren *Malurus cyaneus*

name: Superb Fairy-wren *Malurus cyaneus*
length: 13–14 cm
looks: male blue-and-black except for belly; female light grey-brown, including tail
voice: song a rapid, tremulous trill; loud warning trill; weak contact squeak
habitat: undergrowth of forests, woodlands, heaths, roadsides, farms, gardens; forages on ground and from foliage for insects and small creatures
nesting: spherical, with side entrance; nest made of grass, fine rootlets and bound with webs; nest in low dense shrub, tall grass, bracken
abundance: common

From the earliest days of settlement, birdwatchers noticed that each male fairy-wren had several much plainer fairy-wrens for company. This was especially obvious in spring when each male took on its blue and black breeding plumage, while those that accompanied him remained plain pale grey-brown. Not surprisingly, it was widely assumed that he kept a "harem" of females. "Mormon Wren" became widely accepted as its common name.

Studies in the 1960s revealed a different picture. All members of the group helped rear the young, not just the parents. Far more surprising was the revelation that the group was mostly made up of male birds with just one female: she was partner to the oldest, fully coloured, dominant male. When young birds were fledged, the adults of the group combined forces to drive away all female offspring.

The Superb Fairy-wren builds a small domed nest in a low dense shrub, large clump of grass or bracken. After the three or four eggs hatch, the chicks are fed by most members of the family group.

Noisy Miner *Manorina melanocephala*

name: Noisy Miner *Manorina melanocephala*
length: 25–28 cm
looks: honeyeater with overall pale grey plumage; wings edged yellow, white, barred softly grey around the neck; crown black and face blackish; bare skin behind eye makes a bright yellow and orange triangle; bill bright yellow
voice: many and varied calls, some clear and mellow, quite musical, others loudly complaining or raucous, especially when taken up by the rest of the flock
habitat: open grassy forests and woodlands
nesting: small, rather flimsy bowl of twigs, bark, leaves and cobwebs, softly lined
abundance: common, locally nomadic

These birds have developed an effective social structure and use it to great effect. They live in large, permanently located and rather loosely bonded colonies or communes of up to several hundred birds. These in turn contain small territorial groups of 10–20 birds or more. Birds often gather for communal activities, including feeding, bathing and roosting. The birds of the colony unite to mob and attack any predator in the vicinity, especially snakes and goannas. But these attacks are also used against other bird species, so that local food resources are mostly kept under the control of the colony. In carrying out these strategies the miners use a variety of calls, including one for a ground predator and another for a bird of prey.

Females alone build nests and incubate. Males of the colony then help feed the young. At times there are more than 20 birds helping, in a combined effort, to feed the nestling chicks as often as 30 or more times an hour. This is far more than a single pair could manage, another advantage of this social structure.

Welcome Swallow *Hirundo neoxena*

While swallows and martins are easily recognised as a family, the differences between species are much less obvious. The Welcome Swallow has a glossy black crown and back, becoming dark brown on the wings. Its forehead, face and throat are a bright chestnut.

In the Brisbane region, the Welcome Swallow has adapted to an artificial environment of buildings, bridges and other structures that provide secure and sheltered support for their mud nests. Welcome Swallows are often seen perched in rows along overhead wires, and occasionally on the ground.

Nests seem to exist in the same sheltered parts of buildings and under bridges for decades, and while pairs do return to the same sites, some of those pairs would be reflecting the appreciation of generations of birds for such sheltering homes.

Although the nest is a bowl of hard-dried mud, it is typically very warmly lined with feathers to protect the speckled eggs.

name: Welcome Swallow *Hirundo neoxena*
length: 14–15 cm
looks: typical swallow shape; long pointed wings; deeply forked tail; head and upper back deep glossy black; dark brown on wings and lower back; forehead, sides of neck, throat and upper breast chestnut; breast to undertail white
voice: contact call in flight is a sharp "tzeck" or "tcheck"; song (from a perch) is a mixture of cheery chattering and twittering
habitat: most habitats, but tends to avoid heavy forests and deserts; often around buildings and bridges
nesting: a small semicircular nest of mud, adhered to wall or rock face
abundance: common

Willie Wagtail *Rhipidura leucophrys*

This bird is common and familiar, almost always seen with its long tail fanned wide and restlessly swinging from side to side. Always active and flitting about restlessly, at one moment the Willie Wagtails are snapping insects from the air, and in the next minute are bouncing about after one another on the ground.

In spite of its small size, the Willie Wagtail fearlessly defends its nest territory against any possible threat. This bird will harass even the Australian Hobby, one of the most efficient hunters of small birds and which, if both were out in open airspace together, might well pursue and take the Willie Wagtail. But with the Hobby immobile on its perch, and slow in takeoff, it can do no more than swivel its head about as the annoying Willie Wagtail flutters, abuses and snaps. If ignored, it will strike annoyingly against the raptor's crown or back, before again hovering just out of reach.

Usually the recipient of this unrelenting nuisance gets fed up, and departs the area — just what that bold little bird wanted.

name: Willie Wagtail *Rhipidura leucophrys* length: 19–22 cm
looks: well-known species; identified by constant spreading and wagging of very large tail; head and all upper surfaces, including wings and tail, black; underbody white; narrow white eyebrow
voice: brisk, lively chatter; in contrast, a mellow, clear whistled call, often given repeatedly on moonlit nights "whichity-wheeeit"
habitat: forests and woodlands, margins of rainforests, paperbark swamps, mangroves, parks and gardens of suburbs
nesting: neat, small cup of grass heavily bound with webs and blended onto a fork
abundance: moderately common, may be locally nomadic

Pacific Black Duck *Anas superciliosa*

name: Pacific Black Duck *Anas superciliosa*
length: 48–60 cm
looks: black streaks along crown, through eye, and across lower face
voice: female, typically duck type loud quackings; male a quick "rhaab-rhaab"; in display a high whistle
habitat: diverse wetlands, freshwater and tidal; forages in shallows of lakes, dams, on wet pastures
nesting: large bowl of reeds in dense vegetation, or large hollow of tree
abundance: common

This is Australia's most common duck and also the most widely distributed. While its stronghold is the more consistently wetter coastal country, these ducks quickly become aware of rain in remote semi-desert regions, and promptly move inland to take advantage of the resulting food resources. The Pacific Black Duck is extremely inventive in its choice of feeding habitats, whether it be a farm dam, ornamental fountain, roadside drain or backyard goldfish pond.

This duck is quite large and easily recognised by the buff-white face crossed by a bold blackish line through the eye, a similarly dark crown, and a thinner, dark chin stripe. Except for the colourful iridescent speculum of the wing, the plumage of this bird is dark brown, with buff feather edges.

The Pacific Black Duck's nest may be in a tree hollow or in dense vegetation on the ground. It may be a sparse or thickly walled bowl lined with down. Typically it will hold eight to ten cream-coloured eggs.

Australian Wood Duck *Chenonetta jubata*

name: Australian Wood Duck *Chenonetta jubata*
length: 45–60 cm
looks: large grey duck; male with chestnut head with black crest to nape; both sexes have a heavily spotted breast; in flight, wings white except black outer wings that have dark diagonal stripe
voice: nasal "grouwk", starting low and rising
habitat: grazing rather than truly aquatic duck; feeds on grasslands, but spends some time on water bathing and relaxing; often roosts by water; may use nearby woodlands, pastures, farmlands, urban ponds and dams
nesting: deep hollow of a tree, well lined with down; usually near water
abundance: very common

From a distance this duck appears to have a rather simple plumage — dark brown head, grey body and spotted neck. But when observed up close, or through binoculars, the extensive grey of flanks shows its intricate patterns. Each of the row of large grey feathers covering the flanks is finely cross-barred with dark grey over white.

This duck is primarily a grazer of green pastures where, while mostly feeding upon green herbage, it will also take any insects that are found. It will also take seeds and newly emerging plants of commercial crops such as lucerne and rice.

This duck is abundant and is able to use any small dam in, or close to, its feeding pastures. It is common in outer suburbs where there are generally properties of several hectares or more.

Breeding is from August to January. The Australian Wood Duck's nest is a large hollow trunk or limb of a tree, often near water, but it may be at a distance of a kilometre or more.

Straw-necked Ibis *Threskiornis spinicollis*

So common its plumage is hardly noticed, this ibis has a glossy black back and wings, reflecting an iridescent green to violet sheen. Stiff, straw-like, pale yellow plumes emerge from a ruff around the nape and combine on lower neck and breast as a mass of spiky down-pointing plumes, the origin of the "straw-neck" name.

When travelling, flocks first circle within updrafts, wings held stiffly. On reaching their preferred travelling height they fly in long uneven lines that often form a long "V" formation. Like other ibis, the neck is held straight, the long legs out behind the ends of the stubby tail. Their voice is a harsh, deep croak. The ibis croaks when about to take off; if it is disturbed or alarmed it croaks repeatedly when departing.

Breeding colonies gather and begin courtship and nest-building at any time of the year after rain. Their colonies, often in company with other large waterbirds, are usually found in extensive swamps. Nests are clustered close together on dense reed beds, or on the tops of mangroves or paperbarks in water.

name: Straw-necked Ibis *Threskiornis spinicollis*

length: 60–70 cm

looks: familiar; big black and white ibis; long, downcurved bill; cluster of long, stiff, straw-like plumes at the neck unique to this species

voice: deep croaks, mostly on take-off and landing

habitat: grasslands, wet or dry, especially cultivated and irrigated pastures, wherever there are grasshoppers or similar prey

nesting: usually June–Dec; in colonies often with other large waterbirds; nests crowded together on vegetation standing in water, on bushes, or on low trees; occasionally on the ground

abundance: very common

Australian White Ibis *Threskiornis molucca*

The very long, strongly downcurved bill, together with its large size, immediately identifies this bird as an ibis; the near all-white plumage makes the White Ibis an easily identified species.

The plumage is not quite all-white, with the outermost tips of the wings black. This bird also sports fine, lacy plumes which, on the folded wing, cascade down over the black wingtips and white tail.

The head and upper neck also are black but of bare skin, not plumage. This ibis often feeds in muddy water. When the long bill does not have sufficient reach on its own, the bird's head and half the neck often must go under the water. Consequently, the White Ibis is often to be seen with its plumage stained brown or grey from muddy swamps in which it has been searching for the small underwater creatures that are its natural food.

The Australian White Ibis nests in colonies. The nest is a platform of sticks built on trees, on a trampled platform of reeds, on bare ground, or bushes standing in the water.

name: Australian White Ibis *Threskiornis molucca*

length: 65–75 cm

looks: usual ibis shape; very long downcurved bill; plumage entirely white except black wingtips and plumes; bare skin of head and upper neck black

voice: usually silent except occasional grunt; noisy in breeding colonies

habitat: feeds mostly in shallows of freshwater swamps, lakes, wet pastures, also shallow tidal wetlands; seeks small crustaceans, crabs, frogs, insect larvae

nesting: breeds in closely packed colonies with herons and egrets; builds rough platform in tree standing in water, or on trampled-down clump of rushes

abundance: common to abundant

Laughing Kookaburra *Dacelo novaeguineae*

name: Laughing Kookaburra *Dacelo novaeguineae*
length: 40–47 cm
looks: familiar giant kingfisher of bush and suburbs; big head and massive bill; tiny weak feet; brown above with dull blue patch in wing; underparts and face white, softly barred; dark brown line through eye
voice: famed jovial "laughter", merry chuckling to raucous noise
habitat: gardens, parks, roadsides, woodlands, forests
nesting: large hollow of a tree
abundance: common

This is probably Australia's most widely known and beloved bird — no doubt due to its various outpourings of so-called "laughter". The loud, chuckling call, often starts from a single bird that is then joined by others building up to the full volume of the combined effort, then falling away as individual birds end their part. It is one of the largest birds of the kingfisher family, almost half a metre in length, and with quite a massive bill.

The Laughing Kookaburra is a bird of many habitats. While well known in the suburbs, it is also widespread in many natural vegetation areas of different types, including open forests, woodlands, and partly cleared farmlands. It is a welcome visitor to many backyards and often comes into close contact with adoring humans. Landing on back verandahs and deck railings, these birds will often sing for their supper — cackling heartily until a piece of raw steak or mince is offered. Although it is hard to resist the Laughing Kookaburra's advances, this practice is to be discouraged.

Like most other Australian birds in the kingfisher family, the kookaburra is a terrestrial hunter, patiently waiting for long periods, watching the ground, then gliding down to take any prey when seen. Most prey is small — birds and nestlings, mice and other small mammals. On the other hand, some of their lizard and snake prey are occasionally quite large.

Less well known is Australia's Blue-winged Kookaburra, of northern Australia. While this bird does come south as far as Brisbane, that is almost the southern extremity of its range, so it has a sparse local distribution compared with the Laughing Kookaburra. While it is by far the more colourful in plumage, the Blue-winged Kookaburra's voice would hardly be described as laughter — rather a harsh cackling, fiendish scream, with various loud "klok-klok" noises.

Nesting for the Laughing Kookaburra usually occurs from September to December, but may begin as early as August and extend to February. The nest is usually a large hollow of a tree, or occasionally a hole in an arboreal termite nest. The usual clutch is two to four eggs, which are rather glossy white. These are incubated not only by the female but also by others in the family group.

Sacred Kingfisher *Todiramphus sanctus*

Most owners of a backyard that has been favoured by a pair of nesting Sacred Kingfishers will have noticed that after being absent through winter, their kingfishers return reliably, and noisily, from late August to mid September.

Many Kingfishers in suburban backyards hunt in artificial habitats, taking skinks and spiders from picket fences, frogs from around garden pools, and perhaps even borrowing the occasional goldfish.

While Kingfishers are known for their punctuality and faithfulness to their site, even more amazing is the distance they fly during their time away, and how accurate their navigation is (especially to return to that same backyard and the same nest hollow).

After raising their families through spring and summer, Sacred Kingfishers leave southern Queensland between February and April and fly north. Many cross Torres Strait to New Guinea, Indonesia and even further north. By August these kingfishers are on their return trip, arriving back in their southern breeding territories from late September to early October.

Kingfishers whose homes are backyards, in and around Brisbane, must make their final approach flying over the roads and roofs, homes and commercial centres of our suburbs, often faced with dramatic alterations to the cityscape in their absence. But these small birds, arriving from beyond Australia, seem to know not only that this is their home city but also their home suburb. They know their own backyard, recognise their own tree, even that same small hollow of trunk or branch where last they raised their family, and intend to do so again.

But does each kingfisher pair stay together all the way to coasts and islands far to the north, and return together? Or do they part before travelling north, later to return separately, each remembering months later, that it has a date with another most important small kingfisher in a certain Brisbane backyard, the coming September? Or does the first to arrive at its backyard home just call loudly until it is joined by another in need of a mate?

name: Sacred Kingfisher *Todiramphus sanctus*

length: 20–23 cm

looks: upper parts blue-green; collar buff-white; underbody and underwings white with variable buff tint; buff lores, black through eye

voice: a loud, far-carrying "kik-kik-kik"; near nest "kieer-kieer"

habitat: most open habitats, forests, woodlands; prey of small reptiles, spiders, frogs

nesting: small hollow of tree, usually 3–10 m above ground

abundance: common

Nankeen Kestrel *Falco cenchroides*

name: Nankeen Kestrel *Falco cenchroides*

length: 30–35 cm

looks: very small falcon; cinnamon-rufous on upper surfaces; tail and head grey on male, rufous on female; flight feathers dark grey, barred; tail barred; dark "teardrop" streak down from eye; distinguished from the similar-sized Australian Hobby because it has the ability to hover in one spot (even in gusty winds)

voice: usually silent, but noisy around nest site when breeding

habitat: open grasslands, including unused vacant lots of suburbs, playing fields, road verges, stock paddocks, golf courses, wherever there is grass long enough to hide mice and other prey but not so tall that the overhead bird cannot see down through to most of the ground; natural hunting habitats are open woodlands, grasslands, heathlands, farmlands

nesting: large hollow of a tree, living or dead; eggs laid on bare floor of hollow

abundance: common, but numbers fluctuate with abundance of prey

While the natural habitat of the Nankeen Kestrel consists of the open country of grassy woodlands, heathlands, and any other area with low or sparse ground cover, this species finds an equivalent hunting terrain in the suburbs.

Except the innermost suburbs, Brisbane features many homes on "acreage" lots, which include land for a horse and land that remains unused around or behind a house. Even in the industrial areas are vacant sites with grass, while roadsides, sports fields, river and creek reserves may have grassy land or other low vegetation land where a hovering kestrel might see or flush out small ground birds, mice, spiders and large grasshoppers.

Further out in the Greater Brisbane area are small farm lots, with a few cattle or other livestock. Owners of these properties (with grass as ground cover and a few scattered trees) value the open space and privacy afforded by perhaps 5–10 hectares of land, as do Nankeen Kestrels that hunt in these places.

The Nankeen Kestrel is easily distinguished by its hunting activity. This bird (more than any other) has developed the art of hovering. Whenever this sharp-eyed little raptor spots some suspicious movement in the grass below — it hovers before making a final plunge into the grass.

While hovering it is able to maintain a position relative to the ground, even in the face of gusting winds. The Nankeen Kestrel constantly changes the speed and angle of its wingbeats so that it remains directly above the exact patch of grass where it suspects a mouse or other prey may be hiding.

The Nankeen Kestrel is distinguished by its rufous upper parts, separating it from the Black-shouldered Kite — the only other small bird of prey with similar skill in hovering. Compared with other falcons, its flight is light and wandering, with rapid but comparatively weak wingbeats, much changing of direction, short glides, and pauses in which to hover.

The Nankeen Kestrel is usually silent but quite noisy in the breeding season, with a sharp, rather metallic "ki-ki-ki". Nesting can be in almost any season of the year when there is a lot of prey, but most often between August and December. The nest is made in a large hollow of a tree or cliff, where three to four brown-speckled and blotched eggs are laid.

Bar-shouldered Dove *Geopelia humeralis*

This bird is easily recognised by the bright rufous on the back of its neck and shoulders. Each feather is edged in black, giving the appearance of broken and rather jumbled cross-barring. The throat and upper breast are soft blue-grey in contrast to the warm brown of the back.

In flight this dove travels swiftly, direct and low, with quite a loud whistling sound from the wings that, on the uplift, show off the extensive cinnamon colouring of the bird's underwing plumage.

These doves feed only on the ground, several birds often gathering together, pecking among leaf litter and soil for seeds and the bulbs of sedges and other plants. The doves always keep quite close to water and under the dense vegetation of forests, woodland undergrowth, scrubby margins of swamps, mangroves and creeks, crops and plantations, lantana thickets, parks and gardens with areas of dense shrubbery.

The nest is a frail platform of sticks placed in a shrub, vine thicket or tree, usually within 5 m of the ground.

name: Bar-shouldered Dove *Geopelia humeralis*
length: 27–30cm
looks: medium sized dove; back of neck and shoulders rufous; back and wing brown with heavy black cross-bars; head to breast pale blue-grey, eye and feet deep pink
voice: cheery, double-noted "cookaw, cookaw"
habitat: vicinity of wetlands and in adjoining woodlands, forests, mangroves; forages on ground, mostly for seeds
nesting: small, rough bowl in dense foliage of shrub, or cliff ledge
abundance: common

Peaceful Dove *Geopelia placida (striata)*

The pleasant, clear, lilting call of the Peaceful Dove is one of the characteristic sounds of woodlands, parklands, farmlands and gardens, but only where those habitats are not far from a creek, swamp or dam.

The call can perhaps be expressed as a musical, undulating "coo-wi-ook, coo-wi-ook" given repeatedly, with a throaty, but again pleasantly musical, "quo-r-r-r-r".

These doves are usually seen either in pairs or small groups. When disturbed, all birds will run swiftly, heads bobbing. If put to flight they lift from the ground with rapid wingbeats to dart away, but after quite a short distance will drop to the ground again.

The nesting season begins with a display where the male bows, partly spreads his wings, and widely fans his tail. Although mostly ground birds, Peaceful Doves roost and nest in greater safety in trees. Their flimsy, little, near-flat platform of sticks is placed across a thick horizontal fork of a leafy tree. The two eggs, like those of other doves and pigeons, are white.

name: Peaceful Dove *Geopelia placida (striata)*
length: 20–24 cm
looks: very small slender graceful dove, buffy pale, but extensively barred; large flight feathers and tail dark brown
voice: well-known pleasant lilting "coo-wi-ook, coo-wi-ook", loud and clear
habitat: open forest, woodland
nesting: small platform of sticks on limb, among foliage
abundance: common

Australian Magpie *Gymnorhina tibicen*

name: Australian Magpie *Gymnorhina tibicen*
length: 37–44 cm
looks: plumage boldly black-and-white; straight pointed bill
voice: loud, rich, melodious carolling; great variety of calls and songs
habitat: open woodlands, grasslands, farmlands, gardens and parks
nesting: June–Dec; builds strong, thick-walled bowl of sticks, lined with soft grasses, wool, feathers
abundance: common

One of the country's best-known birds, Australian Magpies live in family or tribal groups that are strongly territorial, defending their own "backyards" against any intruding magpies from another tribe.

The tribe, anywhere from five to twenty birds, needs a huge territory to provide food, especially when there are growing nestlings to be fed. So a tribe of these birds could not find all its requirements in one small backyard, or in some cases even a large lot. These territories cover up 30 hectares and spread over many backyards, streets, and parks. Their fierce defence of territory has earned this bird a reputation for taking on far larger creatures, including humans, near nests.

For the most part, however, magpies are a friendly garden bird, and their gorgeous melodies are a much-loved suburban sound. Semi-tame Australian Magpies are also popular house guests, being fed at barbecues, tables, kitchen windows or doors — friendly, but at the same time bold and proud.

Magpie-lark *Grallina cyanoleuca*

name: Magpie-lark *Grallina cyanoleuca*
length: 26–30 cm
looks: familiar black and white bird; black stripe through eye; white wing patch; tail white with black tip
voice: clear, ringing "tiu-weet, tiu-weet"; in song, female follows on from male alternating in duet
habitat: diverse; woodlands, open country; needs trees and water
nesting: mud bowl on tree limb, heavy and thick-walled, but softly lined
abundance: very common

One of the most common and familiar birds of backyards, the "mudlark" or "peewee" (as the Magpie-lark is also known), has a song that is one of the most delightful of all bird calls, especially when it is a skilfully timed duet between male and female.

Until quite recently these birds were thought to be related to other medium-sized black and white birds. While its common name still includes "magpie" for its superficial resemblance to that bird, later research has shown it to be closely related to the Willie Wagtail and Grey Fantail (all being members of the monarch flycatcher family).

Magpie-larks will usually be present in suburban backyards throughout the year. Pairs strongly defend their nest territory against other birds of their species. Their beautiful synchronised duets are on show during spring months and make them more likely to be noticed during this time. Nesting is typically in spring and early summer, provided rain or other water sources help make the mud from which the nest is built.

Torresian Crow *Corvus orru*

Australia has five species of corvis, two of which live in the Brisbane region. The Torresian Crow is a northern species, with Brisbane near the south-east end of its range. A second corvid, the Australian Raven, extends to Brisbane from the south, the two species overlapping across central and south-eastern Queensland.

In appearance there is usually little difference, but the calls are distinctive. This Torrestian Crow utters an abrupt "uk". In a sequence the notes slow and are further apart — "uk-uk-uk- uk, uk, uk." It also gives a grating, "aark-aark-aark, aaarrgk", ending in deep gurgles. The Australian Raven, however, has first notes of its call much higher, then the sequence descends, eventually fading away in a deep, muffled, gurgling groan — "aaierk, aaaerk, aaergh, aarg-arrgh, arrrururg".

While calling, the Australian Raven has long throat hackles expanded downwards to form an obvious deep shaggy pouch. The Torresian Crow has little if any throat bulge and is smooth, without shaggy, pointed hackles.

name: Torresian Crow *Corvus orru*

length: 48–53 cm

looks: large sleek bird; all black; lacks large shaggy throat hackles of raven; lifts and shuffles wings after landing; white down hidden at base of feathers, which may show when ruffled by wind

voice: short "uk-uk-uk", sometimes slowing at end

habitat: most forested habitats, woodlands, farms, suburbs; takes any small creatures, lizards, nestlings, large insects, carrion

nesting: Aug–Mar or after rain; builds large, rough bowl of sticks, usually high in tree canopy

abundance: common; adults sedentary, juveniles nomadic

Pied Currawong *Strepera graculina*

For Brisbane birdwatchers, the Pied Currawong, unlike the local crows and ravens, presents no great identification difficulty — the two other currawong species living much further south.

Looking something like a mix of crow and magpie, the Pied Currawong is predominantly black, but with conspicuous white patches in the wings, across the base and tip of the tail, and on its undertail coverts. The plumage pattern is most obvious in flight, when white panels are conspicuous on the wings.

Even at a distance, when the plumage details are not visible, the flight is distinctive — a slow, uneven, rather hesitant, rowing wing action, quite unlike the steady strong beat of the crows. The call also is unique — a loud, ringing "curra-currow-currowk".

These are large, noisy birds, using most habitat types and wandering in flocks across quite long distances in autumn and winter. Pied Currawong are common in Brisbane suburbs, parks and gardens.

name: Pied Currawong *Strepera graculina*

length: 42–52 cm

looks: sleek black bird; white on outer wings, tail tip and base; long heavy bill; intense yellow eyes

voice: slow, rollicking "kurrok-kurrow" and "curra-curow -currowk"

habitat: woodlands and forests from coast to alpine; farmlands, gardens

nesting: builds large, rough, but well-lined bowl among slender outer forks of tree canopy, in highest part of tree; 5–25 m high

abundance: abundant, but seasonally nomadic

Rainbow Lorikeet *Trichoglossus haematodus*

name: Rainbow Lorikeet *Trichoglossus haematodus*
length: 26–31 cm
looks: colourful, large lorikeet; bright green back; blue head; deep orange breast; red underwings; greenish nape collar
voice: loud, sharp, vibrant screeches; softer chatterings when feeding
habitat: diverse; rainforests, eucalypt forests, woodlands, heathlands, mangroves, gardens
nesting: small hollow of tree trunk or limb; usually quite high
abundance: common

This bird is the largest of Australia's lorikeets, often seen in screeching flocks that dart across the sky in swift, direct flight — conspicuous, colourful, and very noisy.

While these lorikeets willingly raid flowering trees in suburban backyards, the movement of their flocks will, in most localities, be determined by the availability of natural flowering forests. At times, the large flocks will be absent for considerable periods of time, visiting flowering forests elsewhere. They appear to stay close to sites where food is available most of the year. In the suburbs a greater diversity of planted flowering trees and shrubs, providing food, may increase the number of Rainbow Lorikeets remaining permanently.

Some home owners will undoubtedly prefer not to encourage Rainbow Lorikeets into the backyard — happy that their screeches are heard, but not too often. In large numbers, these birds could make the backyard less attractive to other smaller, less aggressive species such as honeyeaters, smaller lorikeets and the more timid parrots. Coarse mesh over any feeding place reserved for smaller birds should keep these large lorikeets out while allowing smaller species to enter.

In spring, individual pairs break away to find a nest site. A pair choosing a hollow limb above the backyard can bring much life and colour to a home. Whether using a natural hollow or an artificial nest box, they can create much interest with their incredible plumage, chattering, displays and comical antics.

Even just the simple process of preening can be worth watching. It is a captivating sight to behold when a wing is lifted or outstretched and the brilliant red of the underwing linings and the yellow streak across the flight feathers (set against the bright green of back and deep purplish-blue of the head) is displayed. In the backyard so much more can be seen of the birds when compared to a flock hurtling overhead or feeding high in a tree.

Nest boxes, whether intended for these lorikeets or other parrots, may attract a pair to nest in the taller trees of a backyard should no natural hollows be available. Planting trees and large shrubs that provide nectar may benefit these and other nectar-seeking birds.

Sulphur-crested Cockatoo *Cacatua galerita*

A flock of Sulphur-crested Cockatoos, their white plumage set against the deep blue of overhead sky, wheels around the treetops before each bird glides in to land, briefly showing beneath its uplifted wings a delicate tint of yellow. Feeding low in a garden grevillea, destroying its flowers, this bird gives a close view of the plumage that has made its species a world-wide favourite as a cage bird.

After landing, Sulphur-crested Cockatoos may perform various displays — leaning forward and raising the crest, jerking the head up and down and sometimes hanging inverted by the claws or bill with outspread wings showing the yellow tinted underwing. In courtship the male displays with crest raised and head bobbing. Many of these actions are accompanied by loud screeching.

Flocks feed in the mornings, rest in trees through the midday heat, then gather to feed later in the afternoon. Pairs tend to perch close together, but maintain some space, perhaps 0.5–1 m, from other birds. When the flock is feeding on the ground, one bird is usually assigned sentry duty — remaining on watch on a high perch and enabling the rest of the flock to concentrate on feeding. Should the guard sight anything suspicious it will screech an alarm, sending the whole flock into flight.

All these activities make the Sulphur-crested Cockatoo an entertaining as well as a conspicuous and most beautifully plumed species. Much less appreciated is their taste for crops (including wheat, maize, sunflower and canola). Fruit is also damaged, haystacks raided, and bagged grain torn open.

Sulphur-crested Cockatoos are very noisy. Their screech is possibly the loudest of any cockatoo species. They are at their noisiest in the early mornings and late afternoons, especially when departing from and then returning to their roost trees. Their calls are a mix of harsh and ear-piercing screeches, but there are various other grating, whistling and chattering sounds.

The tree hollow chosen as a nest is always large and located in a near-vertical section of tree trunk or major limb. Often a tree near water is chosen or, occasionally, a hole in a cliff.

name: Sulphur-crested Cockatoo *Cacatua galerita*

length: 45–50 cm

looks: very large cockatoo; all white except high, deep yellow crest that curves forward when fully erected or lies back almost flat to show only near back of crown; also pale yellow tint across underwings

voice: noisy; very loud, unpleasant, often piercing screeches and grinding sounds

habitat: diverse; rainforests, wet eucalypt forests, open forests, woodlands, farmlands, coastal mangroves, riverside trees of semiarid regions

nesting: very large hollow of a big tree; eggs laid on wood dust at bottom of deep hollow; occasionally a hole in a cliff

abundance: common to abundant

Bird-friendly Garden Tips

Above: A Yellow-faced Honeyeater and Christmas Bells. When reaching into the flower for nectar the bird's head is thrust deep into the flower, with pollen daubed around head and face, to carry to other bells. Flowers, of different shape, might dust pollen on different parts of the head of each visiting bird.

The most obvious way to attract birds into a garden is to tempt them with food. But this by-passes the natural partnership between birds and flowers. In nature, birds have to work for their nectar. Nectar is only available from each plant for the amount of time it takes birds (and other animals) to help with pollination.

To create a truly bird-friendly garden is to encourage this natural partnership to continue. To put out sugary water or other artificial food is an unnatural handout. It diverts the birds from the pollination work they should be doing in return for their nectar and may also be inappropriate for proper bird nutrition.

Insects are attracted to flowers, and snapped up by the birds. Even lorikeets, while licking nectar from flowers, take in many small insects in those nectar sources, giving the birds a diet far more complete than that provided by sugar-water or bread and honey. When feeding nestlings, many of the honeyeaters bring a constant stream of small insects, rather than nectar. Apparently a varied diet is best for chicks that grow from hatching to adult-size in little more than two weeks.

Bird-pollinated plants range in size from the giant trees of the tall wet eucalypt forests to small ground cover plants (such as Christmas Bells). Birds have evolved in ways that allow them access to the rich source of nectar. Most obvious is the shaping of their bills, tending towards two extremes — the long fine honeyeater bill, or the short broad bill of the lorikeets.

While the spinebill can probe deeply into the flowers of grevilleas, kangaroo-paws and the like, the wide gape of the short, broad bill of the lorikeet is suited to the open-cup flowers of eucalypts that allow its brush-tipped tongue to reach in and take nectar, pollen and insects.

While some of the best bird-attracting flowers are those of shrubs or trees, other low-growing or ground cover plants are more suited for the small garden areas of suburban backyards — kangaroo-paw, Sturt's Desert Pea and the leafless brachysema, for example. One of the best, and most popular among honeyeaters, is *Adenanthos cuneata*, a low shrub with small, deep red flowers surrounded by red leaves. The red leaves that cluster around the flowers are especially bright when backlit by the early morning or late afternoon sun. Among the larger shrubs are many banksias, grevilleas and hakeas. By selecting plants that flower at different times of the year, it is possible to transform the garden into one natural, year-round bird feeder.

Although native flowers and their pollination are the obvious focus of most efforts to establish a bird-friendly garden, there are other bird species that are drawn to backyards for different reasons. Many birds require just the safety of a sheltering, protecting local habitat where insect prey can thrive. The provision of nest boxes where natural hollows are rare will attract species such as the Scaly-breasted Lorikeet.

Fairy-wrens, honeyeaters and most other birds appreciate access to bird baths, well up above reach of dog or cat. A position under dense foliage of tree or tall shrub makes small birds far more comfortable about visiting, as they then have an escape route into foliage, should a raptor attack. Conversely, ground-foraging birds, such as the Magpie-lark and Australian Magpie, are probably more at home on more open patches of lawn where they can forage for insects and grubs.

Suburban lots of just a half hectare or larger might have Rainbow Bee-eaters as residents. Typically these birds prefer an open patch of ground, with sparse or low grass on firm loamy soil (not rock hard, very loose, or waterlogged). The area you choose may be as small as the size of a large room, but preferably not often disturbed or visited, so the birds can feel free.

Moving further out of Brisbane to the larger home blocks of several hectares or more, there are further possibilities, mostly through preserving or replacing native vegetation. An example, one of which will be available to only a small proportion of landowners, is the rehabilitation of watercourses, control of erosion, and improvement of bird habitat. With extra acreage this is possible. A bird species that could directly benefit from this is the Azure Kingfisher.

As with all bird-friendly habitats, anything that could eliminate or discourage feral predators (such as foxes, dogs and cats) will not only improve a bird's chances of survival, but also the chance of it visiting your backyard.

> "Although native flowers and their pollination are the obvious focus of most efforts to establish a bird-friendly garden, there are many other bird species that are drawn to backyards for different reasons."

Below: Eastern Spinebill.

Bird Photography Techniques

Top: A telephoto lens allows a photographer to get close to their quarry while keeping their distance.
Above: Photographing from a hide is a tried and tested method for capturing a bird's natural behaviour.

When photographing birds in their natural habitat, the major difficulty to overcome is birds' fear of people.

Their shyness can be overcome in two ways — by the choice of optimum photographic equipment, and by the employment of techniques planned to ensure the safety of the subject. This has the highest priority. The aim is to ensure that the bird is not aware of the presence of photographer and camera, or to ensure that the bird is so gradually introduced to the photographer and equipment that it will accept their presence without being disturbed from its usual activities. Keeping this in mind, your photos or videos will portray the bird in a natural state.

Because wild birds are usually wary, only the most patient, considerate and cautious approach by the photographer can bring birds close enough to a camera with flash equipment. When working at nests it is most important to proceed slowly and to abandon all attempts at photography should the birds appear disturbed or reluctant to continue with their normal behaviour.

Most birds have now been photographed at their nests. With the coming of equipment that was not available to wildlife photographers until the last decade or so, birds can be photographed at much greater distance, comparable to binocular observation distance. Given that birdwatchers are usually in sizeable (and audible) groups, the lone, silent and almost motionless photographer, camera steady on a tripod, is almost always going to be able, with care, to get closer to their subject without causing much disturbance.

Certainly birdwatching and photography has only the most minute impact if proper care is taken. For example, waders and other shoreline birds are far more often put to flight by vehicles, dogs, joggers, walkers, and other users, than by observers or photographers of birds.

Participants of jogging, walking and vehicular activities greatly outnumber the few taking some knowing interest in the birds, probably by thousands-to-one ratio. The multitude of beach users mostly seem quite unaware of the presence of shorebirds, nor that there may be near-invisible nests on the sand. The occasional presence of someone watching or photographing can, on occasion, result in harmful activities being diverted, suspended, or warned, or damaging activities reported.

Decades ago there was no distant, macro or zoom, photo option and birds had to be photographed close using hides at waterholes, near carrion, nests and any other place where birds would return or might be likely to go. A camouflaged hide, securely tied against wind flap or movement and already in position days earlier (and accepted by birds as safe), is one of the least disturbing ways to get close to a bird for observations or photography. Indeed, hides are a feature of many national park boardwalks used by groups rather than a solitary silent photographer. Today the emphasis is on photography from afar, using huge telephoto lenses, seeking authentic and natural behaviour rather than portraits of birds.

For most amateur bird photographers one of the many better quality 35mm single lens reflex cameras will be a suitable choice. It is important that the camera has a quiet shutter, and that a long telephoto lens is available. Today one makes the decision also, whether to use film or digital equipment. The quality of result from the best digital cameras, and even the mid-priced models is excellent, coupled with convenience and low-cost shooting. Add to this the ability to safeguard against loss of hard-earned images by easily making identical copies, plus the value of digital correction of images, and the digital path looks ever more attractive. Of course images on film can be scanned, but for high quality, publishable scans, there is additional expense.

For those who already have a large, and expensive investment in equipment, such as big telephoto lenses, the choice to change media could be daunting. However, some manufacturers now make digital camera bodies to fit most of their expensive lenses previously manufactured for their film camera bodies. So it is unlikely that existing investment in costly lenses will be wasted. Unfortunately, the same does not apply to the film camera bodies, if one decides to go entirely digital. For many photographers, perhaps both types will continue side by side for some time.

Photography plays a huge part in maintaining or building public interest and information for conservation purposes. Certainly photography, both still and video, will continue to be a major contributor to building or maintaining public interest, and thereby the political will, to ensure all environmental aims can be realised.

Above: Rufous Night Heron.

“Photography still plays a huge part in maintaining or building public interest and information for conservation purposes.”

Birds of Brisbane & Surrounds

This book covers Brisbane and its environs — about the limit of a one-day excursion. Unfortunately, no book or printed information ever remains up to date — walking tracks and even road access may change. For this purpose, websites often provide the best information (provided they are updated regularly). Park rangers will also be able to give helpful information on the day of any planned excursion through a national park.

South-eastern Queensland has two birding groups, each with their own website containing detailed, up-to-date information. Two of the best websites to consult are:

Birds Australia Southern Queensland (BASQ):
http://www.users.bigpond.com/basqld/
BASQ is the local branch of the national Birds Australia. Members include amateur birdwatchers and professional bird observers interested in native birds. Activities include bird counts and surveys (gardens, farms and on stock routes); atlassing and campouts. Facts sheets are published that cover, among other topics, feeding wild birds; threatened species, bird flu and other avian data.
http://www.users.bigpond.com/basqld/

Birds Queensland:
http://www.birdsqueensland.org.au/
The Birds Queensland website reflects the work of the Queensland Ornithological Society Inc. It provides a huge resource of information dedicated to the conservation and scientific study of birds. It outlines places to visit, monthly meetings, rare and unusual bird alerts and info sheets. Past excursions by the group are also recorded.

Between each of these groups there are a number of classes for bird enthusiasts to attend. Topics for these classes include information on bird identification, use of field guide books, choice and use of binoculars, telescopes and cameras and other practical tips. Field excursions close to the city and campouts to more distant sites let beginners meet and learn from experienced bird observers. This is the best way to get to know the region's birds and familiarise yourself with the techniques and equipment that give the best results.

Especially valuable is the website "Gazetteer" (http://birdsqueensland.org.au/bq-gazetteer-oth.html) that Birds Queensland has produced. This names most of the top birdwatching sites for both south-eastern Queensland and the rest of Queensland. Each site profile gives the name of the site, the access route, the tracks or trails through the reserve, the landscape, geology, vegetation and facilities.

Birds Queensland has also, with funding from the Brisbane City Council, published *Bird Places of Brisbane*, subtitled *Where to Find Native Birds in Brisbane.* This has a map of Brisbane and its suburbs, outlining 48 of the best birding locations.

Brisbane City Council Track Maps and Bird Lists are available at Boondall and Downfall Creek (Raven St Reserve) Visitor Centres, Council Business Offices, and some Council Libraries.

Birds Australia, the national parent body of Birds Australia Southern Queensland, is the national voice of bird observers. Publications include the monthly *Wingspan*, with features of general birding interest, supported by excellent illustrations and the journal *The Emu*, of more scientific interest. Phone 03 9882 2622, send an email to mail@birdsaustralia.com.au or visit their website at www.birdsaustralia.com.au

Bird Books

Morcombe, Michael 2003, *Michael Morcombe Field Guide to Australian Birds*, Steve Parish Publishing, Brisbane. Further information at www.michaelmorcombe.com.au and also the publisher's website at www.steveparish.com.au

Poole, Stephen 1996, *Wild Places of Greater Brisbane*, Queensland Museum, South Brisbane.

Noyce, John (ed.) *A Bird Watcher's Guide to Redcliffe, Pine Rivers and Caboolture*, Wildlife Preservation Society of Caboolture Shire Inc., Brisbane.

Parish, Steve 2003, *Photograph Australia with Steve Parish*, Steve Parish Publishing, Brisbane.

Wieneke, Jo 2000, *Where to Find Birds in North-East Queensland*, Jo Wieneke, Belgian Gardens.

Thomas, Richard and Thomas, Sarah 1996 *The Complete Guide to Finding the Birds of Australia*, Frogmouth Publications, Oakington, UK.

Bird Songs, on CD, David Stewart, Naturesound. Visit www.steveparish.com.au for the many CD collections of bird calls and songs.

The description and information of birdwatch walks in this book are intended only as a general indication and should not be used as the sole guide to walking. It is the responsibility of the user to have detailed and accurate maps, equipment and provisions when undertaking any of the suggested birdwatching activities, and to obtain the advice of rangers or landowners regarding safety, current conditions, and other relevant information. When in doubt, walkers should keep to made roads or trails, and never walk alone or without advising someone of routes and times. Unknown swamps and lakes should not be entered alone, and caution should be taken in case of deep mud, deep water or other hidden hazards. Bushwalkers should obtain the permission of owners or occupiers of private land likely to be traversed, observe all requirements and legal obligations regarding camping or cooking fires, and leave all gates as found.*

Index

Index